Ethical Diversity in Healthcare Delivery

Ethical Diversity in Healthcare Delivery

Healthcare Ethics & Beyond

Adam Musah, PhD

Health Management and Policy Specialist

Library of Congress Control Number:		2017900851
ISBN:	Hardcover	978-1-5245-7733-9
	Softcover	978-1-5245-7732-2
	eBook	978-1-5245-7734-6

To order additional copies of this book, contact:
Xlibris
1-888-795-4274
www.Xlibris.com
Orders@Xlibris.com
741982

Praise for the "Ethical Diversity In Healthcare Delivery"

"Making our world a better place to live – sure, we can! Dr. Musah has done it again. In the field of healthcare, the writer calls on health providers to be aware of patient sensitivities and to also learn and appreciate our human diversity, in creed, color and culture. To achieve a lasting solution in every human relationship, the collaborative process is necessary and that can only be achieved through trust, which is the result of genuine qualities of compassion and caring.

As an Imam, I believe this book will be a healthy tool not only to healthcare workers but also for all who care about the survival of humanity in the age of diversity. Society is seriously weakened when a patient is denied better care due to their creed, color, culture and other demarcations. To care is a universal principle. This book is a timely subject providing us with important life values. Congratulation Dr. Musah for your service to humanity."

— Abdul-Rahman Yaki,

Imam of the Islamic Center of the Capital District

(Albany, NY)

"A vital guide for a healthcare access and utilization in minority communities"

_ Alhaji Baba Allaino

A minority Community Leader (Bronx, NY)

"A concise must-read primer for all caregivers in a culturally divergent, yet fast converging world."

__ Samory Cisse Ali-Danbukar, *Esq. (Bronx, NY)*

"As a Biostatistician, I have had a good chance to look over the fine work in this new book …. My general conclusion is that it is extremely a good read for both students and healthcare professionals on different levels and belief systems."

__ Sheiba I. Mas-Oud,

Associate Professor – QCC, Research Assistant – UMASSMED

(-----, MA)

Dedication

ONE SHOULD BE thankful to Almighty God at all times for the strength and wisdom in accomplishing tasks, and or successfully going through daily hurdles, and for the enormous academic achievements in addition to writing this book. Optimistically, this book will be helpful to healthcare providers and students in the healthcare discipline. Coming up with this book and my educational accomplishments in general would not have been possible without the love and patience of my family. My lovely family, to whom this book is dedicated, has been a constant source of love, moral and material support during the long years of my education. It is also dedicated to my beloved parents and other family members who have returned to Almighty God. May He have mercy on them and all mankind!

ABSTRACT

Breadth

The objective of this Knowledge Area Module (KAM) is to identify, explore, and articulate the principles and assumptions behind contemporary issues and the ethical diversity in the delivery of health services. The breadth component focuses on health services delivery and ethical diversity in health care provision. Ethical topics addressed include: Ethics in health services management, theory of justice, and introduction to ethical theory. These publications explained the concept surrounding ethical dilemma in health care. Furthermore, I reviewed the following literature for further clarity on the subject matter: The American Counseling Association ethical approach, the code of ethics of the American Rheumatology which shed light on designing and supporting college's mission, the code of ethics and professional responsibilities for healthcare ethics consultants, and ethical issues in community health services.

ABSTRACT

Depth

The Depth section focuses on health services delivery and ethical decision-making in communities and other health entities. I reviewed 14 bibliographies on current health research findings. Some of the issues that triggered probing were: Assessment of harm potential and factors influencing patients in participation decision-making, reducing premature infants' length of stay, and improving patient's mental health outcome, parental involvement in treatment decisions regarding their critical ill children, and Informed consent and malpractice lawsuits plaguing healthcare organizations. Some of the findings laid strong foundation for further research while others provided recommendations for rectifying ethical dilemma in health services delivery.

ABSTRACT

Application

The application component focuses on ethical theories applied in health services delivery. I applied the concept explored in the breath components and the analysis of findings on current research in the depth component and provided guidelines and recommendations. Adhering to these recommendations and or guidelines will help alleviate several ethical problems in the healthcare industry. Additionally, this project entails the critical review of the following publication: Caring for the dying patient, Islamic medical ethics and the Islamic Law, national and international approaches to ethical guidelines implementations and periodic evaluations, general health education on doctor and patient responsibilities and informed consent general consideration.

Table of Contents

Ethical Diversity Issues In Health Services Delivery

INTRODUCTION

IN THE WAKE of massive technological innovation and increased patient diversity, the most daunting challenges facing healthcare industry is the provision of medical care within the mosaic of ethical diversity (Genius and Lipp, 2013). Ethics is relevant to all aspects of healthcare delivery system in addition to biomedical research that involves human subjects in the prospect of discovering new ways of benefiting people's health. Consequently, central to the process of health administration is the management of conflicting values, and the rapid changes that occur in our societies and health services field have made the task more challenging to manage. Therefore, all medical professionals and health researchers are required to undergo formal training in the theories and principles of ethics in healthcare (Genius and Lipp, 2013).

To clearly understand ethical issues and the strategies for managing them, I reviewed the following theories and models: Ethics in Health Services Management by Darr (1997), a Theory of Justice by Rawls (1971), and Introduction to Ethical Theory by Rogerson (1991). Additionally, I reviewed the following contemporary classical works on theories and general articles on ethical knowledge for in depth knowledge on ethics in healthcare: Group Counseling – Ethics and Professional Issues by Forester-Miller and Rubenstein, Making Hard Choices – Clarifying Controversial ethical issues, and A Practitioner's Guide to Mental Health Ethics by Haas and Malouf (1989).

This knowledge area module is designed for assistance in developing competence in the application of ethical theories and principles to decision making related to the provision and delivery of preventive, diagnostic, curative, and restorative level of health services to selected population. In other words, knowledge derived from selected readings will be helpful in strategizing an ethical management tool for healthcare organizations.

Breadth

Ethical Diversity Issues in Health Services Delivery

THE FOLLOWING THEORIES and models were reviewed to clarify various concepts shaping the ethical dilemma in the delivery of health services in the United States of America and other parts of the world. The classical works of Rawls, Joseph Fletcher and Immanuel Kant were compared and contrasted to clarify whether additional steps are required by our health care providers to address ethical concerns effectively. Some of these concerns may be about the impact of religion and cultural values on informed consent. Further, the issues of morality in ethical principles of health institutions were scrutinized to see if they impact one another.

Ethical Theories

If the practitioners of bioethics do not rely solely on cultural norms and emotions, what are their sources of determining what is right or wrong? The most comprehensive source is a theory of ethics, a broad set of moral principles or perhaps just one overriding principle that is used in measuring human conduct. Divine law is one such source, of course, but even in the Western religious traditions of bioethics both the Jewish and Catholic religions have rich and comprehensive commentaries on ethical issues, and the Protestant religion has a less cohesive but still important tradition, the law of God is interpreted in terms of human moral principles. Recently, bioethicists have paid more attention to analyzing the teachings of other religious traditions, such as Islam, Buddhism, Confucianism, and other eastern religions. A theory of ethics must be acceptable to many groups, not just the follower of one religious tradition. Most writers outside the religious traditions and some within them have looked to one of three major traditions in ethics: teleological theories, deontological theories, and natural law theories (Rawls, 1971).

Teleological Theories

Teleological theories are based on the idea that the end or purpose (from the Greek *telos,* or end) of the action determines its rightness or wrongness. The most prominent teleological theory is utilitarianism. In its simplest formulation, an act is moral if it brings more good consequences than bad ones. Utilitarian theories are derived from the works of two English philosophers: Jeremy Bentham (1748-1832) and John Stuart Mill (1806-1873 and Rawls, 1971). Rejecting the absolutist religious morality of his time, Bentham proposed that "utility" the greatest good for the greatest number-should guide the actions of human beings. Invoking the hedonistic philosophy of Epicurean Greeks, Bentham said that pleasure *(hedon* in Greek) is good and pain is bad. Therefore, actions are right if they promote more pleasure than pain and wrong if they promote more pain than pleasure. Mill found the highest utility in "happiness," rather than pleasure. Mill's philosophy is echoed in the Declaration of Independence's espousal of "life, liberty, and the pursuit of happiness." Other utilitarian have looked to a range of utilities, or goods including friendship, love, devotion, and the like that they believe ought to be weighed in the balance to the utilitarian calculus.

Utilitarianism has a pragmatic appeal. It is flexible, and it seems impartial.

However, its critics point out that utilitarianism can be used to justify suppression of individual rights for the good of society "the ends justify the means" and that it is difficult to quantify and compare "utilities," however they are defined. Utilitarianism, in its many forms, has had a powerful influence on bioethical discussion, partly because it is the closest to the case-by-case risk/benefit ratio that physicians use in clinical decision-making. Joseph Fletcher, a Protestant theologian who was one of the pioneers in bioethics in the 1950s, developed utilitarian theories that he called *situation ethics*. He argued that a true Christian morality does not blindly follow moral rules but acts from love and sensitivity to the particular situation and the needs of those involved. He has enthusiastically supported most modern technologies on the grounds that they lead to good ends (Rawls, 1971). According to Rogerson (1991), writers in this volume who use utilitarian theories to arrive at their moral judgments include Lawrence O. Gostin, who defends giving public health agencies sweeping powers in a bioterrorist threat; and Jerod M. Loeb and his colleagues, who defend animal experimentation.

Deontological Theories

The second major type of ethical theory is deontological. The rightness or wrongness of an act, these theories hold, should be judged on whether or not it conforms to a moral principle or rule, not on whether it leads to good or bad consequences (Rogerson, 1991). The primary exponent of a deontological theory was Immanuel Kant (1724-1804), a German philosopher Kant declared that there is an ultimate norm, or supreme duty, which he called the "Moral Law." He held that an act is moral only if it springs from a "good will," the only thing that is good without qualification.

We must do good things, said Kant, because we have a duty to do them, not because they result in good consequences or because they give us pleasure although that can happen as well. Kant constructed a formal "Categorical Imperative," the ultimate test of morality:

Recognizing that this formulation was far from clear, Kant said the same thing in three other ways. He explained that a moral rule must be one that can serve as a guide for everyone's conduct; it must be one that permits people to treat each other as ends in themselves, not solely as means to another's ends; and it must be one that each person can impose on himself by his own will, not one that is solely imposed by the state, one's parents, or God. Kant's Categorical Imperative, in the simplest terms, says that all persons have equal moral worth and that no rule can be moral unless all people can apply it autonomously to all other human beings. Although on its own, Kant's Categorical Imperative is merely a formal statement with no moral content at all, he gave some examples of what he meant: "Do not commit suicide," and "Help others in distress." (Rogerson K. 1991).

Kantian ethics is criticized by many who note that Kant gives little guidance on what to do when ethical principles conflict, as they often do. Moreover, they say, his emphasis on autonomous decision-making and individual will neglects the social and communal context in which people live and make decisions. It leads to isolation and unreality. These criticisms notwithstanding, Kantian ethics has stimulated much current thinking in bioethics. In this volume, the idea that certain actions are in and of themselves right or wrong underlies, for example, Patrick Lee and Robert P. George's argument against abortion because it involves killing a human being; Tom Regan's opposition to animal research; and President's Council on

Bioethics' opposition to federal funding of human stem cell research (Rogerson, 1991).

According to (Rosenbaun, 1982), two modern deontological theorists are philosophers John Rawls and Robert M. Veatch. In A Theory of Justice (1971), Rawls places the highest value on equitable distribution of society's resources. He believes that society has a fundamental obligation to correct the inequalities of historical circumstance and natural endowment of its least well off members. According to this theory, some action is good only if it benefits the least well off. It can also benefit others, but that is secondary. His social justice theory has influenced bioethical writings concerning the allocation of scarce resources. Veatch has applied Rawlsian principles to medical ethics. In his book, A Theory of Medical Ethics (1981), he offers a model of social contract among professionals, patients, and society that emphasizes mutual respect and responsibilities. This contract model will, he hopes, avoid the narrowness of professional codes of ethics and the generalities and ambiguities of more broadly based ethical theories (Rosenbaun, 1982).

Natural Law Theory

The third strain of ethical theory that is prominent in bioethics is natural law theory, first developed by St. Thomas Aquinas (1223-1274). According to this theory, actions are morally right if they accord with our nature as human beings. The attribute that is distinctively human is the ability to reason and to exercise intelligence. Thus, argues this theory, we can know the good, which is objective and can be learned through reason. References to natural law theory are prominent in the works of Catholic theologians and writers; they see natural law as ultimately derived from God but knowable through the efforts of human beings. The influence of natural law theory can be seen in the issues on human stem cell research and genetic enhancement, declared Thomas Aquinas (Rogerson, 1991).

Theory of Virtue

The theory of virtue, another ethical theory with deep roots in the Aristotelian tradition, has recently been revived in bioethics. This theory stresses not the morality of any particular actions or rules but the disposition of individuals to act morally, to be virtuous. In its modern version, its primary exponent is Alasdair MacIntyre, whose book After Virtue (1980) urges a return to the Aristotelian model. Gregory Pence has applied the theory of virtue directly to medicine in Ethical Options in Medicine (1980); he lists temperance in personal life, compassion for the suffering patient, professional competence, justice, honesty, courage, and practical judgment as the virtues that are most desirable in physicians. Although this theory has not yet been as fully developed in bioethics as the utilitarian or deontological theories, it is likely to have particular appeal for physicians, many of whom have resisted formal ethics education on the grounds that moral character is the critical factor and that one can best learn to be a moral physician by emulating one's mentors (Berg and Braddock, 2001).

Although various authors, in this volume and elsewhere, appeal in rather direct ways to either utilitarian or deontological theories, often the various types are combined. One may argue that both particular actions are immoral in and of itself and that it will have bad consequences, some commentators say even Kant used this argument. In fact, probably no single ethical theory is adequate to deal with ramifications of all the issues. In that case we can turn to a middle level of ethical discussion. Between the abstractions of ethical theories, Kant's Categorical Imperative, and the specifics of moral judgment are a range of ethical principles concepts that can be applied to particular cases (Berg and Braddock, 2001).

Ethical Principles

In its four years of deliberation in the 1970s, the National Commission for the Protection of Human Subjects of Biomedical and Behavioral Research grappled with some of the most difficult issues facing researchers and society: When, if ever, is it ethical to do research on fetuses, on children, or on people in mental institutions? This commission, which was composed of people from various religious backgrounds, professions, and social strata, was finally able to agree on specific recommendations on these questions, but only after they had finished their work did the commissioners try to determine what ethical principles they had used in reaching a consensus. In their

Belmont Report (1978), named after the conference center where they met to discuss this question, the commissioners outlined that respect for persons, beneficence, and justice are the three items that should govern the conduct of research with human beings. These three principles, they believed, are generally accepted in our cultural tradition and can serve as basic justifications for the many particular ethical prescriptions and evaluations of human action. Because of the principles' general acceptance and widespread applicability, they are at the basis of most bioethical discussion. Although philosophers argue about whether other principles preventing harm to others or loyalty, for example, ought to be accorded equal weight with these three or should be included under another umbrella, they agree that these principles are fundamental (Belmont Report 1978).

Respect for Persons

Respect for persons incorporates at least two basic ethical convictions, according to the Belmont Report. Individuals should be treated as autonomous agents, and persons with diminished autonomy are entitled to protection. The derivation from Kant is clear. Because human beings have the capacity for rational action and moral choice, they have a value independent of anything that they can do or provide to others. Therefore, they should be treated in a way that respects their independent choices and judgments. Respecting autonomy means giving weight to autonomous persons' considered opinions and choices, and refraining from interfering with their choices unless those choices are clearly detrimental to others. However, since the capacity for autonomy varies with age, mental disability, or other circumstances, those people whose autonomy is diminished must be protected - but only in ways that serve their interests and do not interfere with the level of autonomy that they do possess (Rawls, 1971).

Two important moral rules are derived from the ethical principle of respect for persons: informed consent and truth telling. Persons can exercise autonomy only when they have been fully informed about the range of options open to them, and the process of informed consent is generally considered to include the elements of information, comprehension, and voluntariness. Thus, a person can give informed consent to some medical procedure only if he or she has full information about the risks and benefits, understands them, and agrees voluntarily that is, without being coerced or pressured into agreement. Although the principle of informed consent has become an accepted moral rule and a legal one as well, it is difficult if not impossible-to achieve in a real-world setting. It can easily be turned into a legalistic parody or avoided altogether. But as a moral ideal, it serves to balance the unequal power of the physician and patient. (Parker, 2007).

According to SophiaOmni (2012), truth telling is another important moral ideal derived from the principle of respect for persons.

On the Supposed Right to Lie From the Benevolent
Motives: Kant stated that it is a duty to tell the truth.
The notion of duty is inseparable from the notion of
right. A duty is what in one being corresponds to the
right of another. Where there are no rights there are
no duties, but only towards him who has a right to the
truth. But no man has a right to a truth that injures
others. To tell the truth is a duty, but only towards him
who has a right to the truth (p. 1).

Justice

The third ethical principle that is generally accepted is justice,
which means, "what is fair" or "what is deserved." The Belmont Report
of 1979 indicated that an injustice occurs when some benefit to which
a person is entitled is denied without good reason or when some bur-
den is imposed unduly, and that the interpretation should be that
equals should be treated equally. However, some distinctions - such
as age, experience, competence, physical condition, and the like -
can justify unequal treatment. Those who appeal to the principle
of justice are most concerned about which distinctions can be made
legitimately and which ones cannot.

According to Rawls (1971), one important derivative of the
principle of justice is the recent emphasis on "rights" in bioethics.
Given the successes in the 1960s and 1970s of civil rights movements
in the courts and political arena, it is easy to understand the appeal of

"rights talk." An emphasis on individual rights is part of the American tradition, in a way that emphasis on the "common good" is not. The language of rights has been prominent in the abortion debate, for instance, where the "right to life" has been pitted against the "right to privacy" or the "right to control one's body." The "right to health care" is a potent rallying cry, though it is one that is difficult to enforce legally. Although claims to rights may be effective in marshaling political support and in emphasizing moral ideals, those rights may not be the most effective way to solve ethical dilemmas. Our society, as philosopher Ruth Macklin has pointed out, has not yet agreed on a theory of justice in health care that will determine who has what kinds of rights and-the other side of the coin-who has the obligation to fulfill them (Rawls, 1971).

Analysis of Ethical Theories

These three fundamental ethical principles of respect for persons, beneficence, and justice will carry weight in ethical decision-making. But what happens when they conflict? On each side of the issues are writers who appeal, explicitly or implicitly, to one or more of these principles. For example, Jean Toal sees beneficence as paramount, and she would criminalize drug-using behavior by pregnant women in order to prevent harm to their fetuses. Lynn M. Paltrow finds such a policy unjust because it singles out certain risks and certain women for state intervention. Some of the issues are concerned with how to interpret a particular principle: Whether, for example, it is more or less beneficent to allow a physician to assist in suicide, or whether society's interest in obtaining transplantable organs for those who need them and allowing payment for them (Rosenbaum, 1982).

Will it ever be possible to resolve such fundamental divisions are those that are not merely matters of procedure or interpretation but of fundamental differences in principle? Lest the situation seem hopeless, consider that some consensus does seem to have been reached on questions that seemed equally tangled a few decades ago. The idea that government should play a role in regulating human subjects research was hotly debated, but it is now generally accepted - at least if the research is medical, not social or behavioral in nature, and is federally funded. Moreover, the appropriateness of using the criterion of brain death for determining the death of a person and the possibility of subsequent removal of their organs for transplantation has largely been accepted and written into state laws. The idea that a hopelessly ill patient has the legal and moral right to refuse treatment that will only postpone dying is also well established though it is often hard to exercise because hospitals and physicians continue to resist it. Finally, nearly everyone now agrees that health care is distributed unjustly in the United States - a radical idea only a few years ago. There is, of course, sharp disagreement about whose responsibility it is to rectify the situation, the government or the private sectors? (Ruth and Beauchamp, 1986).

Besides the virtue theory, already described, two other candidates have their defenders. The ethics of caring has been presented as an alternative to traditional bioethics reasoning. Women, it is claimed, embody an ethic of caring, which is itself a prime aim of healing relationships. An ethic of caring would focus on relationships rather than autonomy, on reconciliation rather than winning an argument, and on nurturing rather than imposing dominance (Forester-Miller and Rubinstein, 1992). While the absence of caring relationships is clearly a problem in modern health care, this view has been severely criticized by many, including women, as failing to provide a sufficient basis for replacing ethical principles (Forester-Miller and Rubinstein, 1992).

A final form of analysis is clinical ethics. Its practitioners focus on the clinical realities of moral choices as they emerge in ordinary health care. It is not antithetical to principles but brings abstractions back to reality by measuring proposed solutions against the real world in which doctors and patients live and work. The refusal of treatment on the grounds of futility builds on clinical ethics and real cases. Edmund Pellegrino, a distinguished physician and ethicist, has seen many changes in more than 50 years he has been involved in medicine. Looking toward the future, he does not see the death of principles, but he does foresee some changes. Physicians and other health workers must become familiar with shifts in contemporary moral philosophy if they are to maintain a hand in restructuring the ethics of their profession (Forester-Miller and Rubinstein, 1992).

Medicine and Moral Arguments

Carol Levine narrated this story that in the fall of 1975 a 21-year-old woman lay in a New Jersey hospital-as she had for months in a coma, the victim of a toxic combination of barbiturates and alcohol. Doctors agreed that her brain was irreversibly damaged and that she would never recover. Her parents, after anguished consultation with their priest, asked the doctors and hospital to disconnect the respirator that was artificially maintaining their daughter's life. When the doctors and hospital refused, the parents petitioned the court to make her legal guardian so that they could authorize the withdrawal of treatment. After hearing all the arguments, the court sided with the parents, and the respirator was removed. Contrary to everyone's expectations, however, the young woman did not die but began to breathe on her own, perhaps because, in anticipation of the court order, the nursing staff had gradually weaned her from total dependence on the respirator. She lived for 10 years until her death in June 1985 - comatose, lying in a fetal position, and fed with tubes - in a New Jersey nursing home (Levine, 2008).

The young woman's name was Karen Ann Quinlan, and her case brought national attention to the thorny ethical questions raised by modern medical technology: When, if ever, should life-sustaining technology be withdrawn? Is the sanctity of life an absolute value? What kinds of treatment are really beneficial to a patient in a "chronic vegetative state" like Karen's? And, perhaps the most troubling question, who shall decide? These and similar questions are at the heart of the growing field of biomedical ethics, or as it is usually called *bioethics.*

Carol Levine further stated that ethical dilemmas in medicine are, of course, nothing new. They have been recognized and discussed in Western medicine since a small group of physicians led by Hippocrates on the Isle of Cas in Greece, around the fourth century B.c., subscribed to a code of practice that newly graduated physicians still swear to uphold today. But unlike earlier times, when physicians and scientists had only limited abilities to change the course of disease, today they can intervene in profound ways in the most fundamental processes of life and death. Moreover, ethical dilemmas in medicine are no longer considered the sale province of professionals. Professional codes

of ethics, to be sure, offer some guidance, but they are usually unclear and ambiguous about what to do in specific situations. More important, these codes assume that whatever decision is to be made is up to the professional, not the patient. Today, to an ever-greater degree, lay-people patients, families, lawyers, clergy and others want to and have become involved in ethical decision making not only in individual cases, such as the Quinlan case, but also in large societal decisions, such as how to allocate scarce medical resources, including high technology machinery, newborn intensive care units, and the expertise of physicians. While questions about the physician-patient relationship and individual cases are still prominent in bioethics today, the field covers a broad range of other decisions as well, such as the use of reproductive technology, the harvesting and transplantation of organs, equity in access to health care, and the future of animal experimentation (Levine, 2008).

This involvement is part of broader social trends: a general disenchantment with the authority of all professionals and, hence, a greater readiness to challenge the traditional belief that "doctor knows best"; the growth of various civil rights movements among women, the aged, and minorities of which the patients' rights movement is a spin-off; the enormous size and complexity of the health care delivery system, in which patients and families often feel alienated from the professional; the increasing cost of medical care, much of it at public expense; and the growth of the "medical model," in which conditions that used to be considered outside the scope of physicians' control, such as alcoholism and behavioral problems, have come to be considered diseases (Levine, 2008).

Bioethics began in the 1950s as an intellectual movement among a small group of physicians and theologians who started to examine the questions raised by the new medical technologies that were starting to emerge as the result of the heavy expenditure of public funds in medical research after World War II. They were soon joined by a number of philosophers who had become disillusioned with what they saw as the arid abstractions of much analytic philosophy at the time and by lawyers who sought to find principles in the law that would guide ethical decision making or, if such principles were not there, to develop them by case law and legislation or regulation. Although these four disciplines - medicine, theology, philosophy, and law still dominate the field, today bioethics is an interdisciplinary effort, with political scientists, economists, sociologists, anthropologists, nurses, allied health professionals, policymakers, psychologists, and others contributing their special perspectives to the ongoing debates (Forester-Miller and Rubenstein, 1992). The issues that are currently discussed attest to the wide range of bioethical dilemmas, their complexity, and the passion they arouse. But if bioethics today is at the frontiers of scientific knowledge, it is also a field with ancient roots. It goes back to the most basic questions of human life: What is right? What is wrong? How should people act toward others? And why? (Levine, 2008).

Ethics Code

I reviewed the Code of Ethics of the American college of Rheumatology designed to guide the college throughout its educational program. I also reviewed other organizations' codes for further understanding and application. The College successfully maintained its reputation in the scientific and medical communities and with the general public as a result of its well designed code of ethics which ensures that a certain type of standard or reputation is maintained with regards to ethical issues stemming from daily professional activities faced by its members, and which directly affects ethical matters on the College's scientific and educational mission according to the American Counseling Association (2005).

Medical Ethics General Principles

The general principles of ethics is the first part of this code of ethics of the generally accepted standards of professional conduct that guides members of various institutions. The list below is the American College of Rheumatology's principles of professional conducts guiding members in relationship with patients and the public (ACR, 2015). The list ranges from competent medical care with compassion and respect, honesty in all professional interactions, protecting the confidentiality of the physician/patient relationship, continuing study and improve on scientific knowledge, disclosing new medical innovations to patients and colleagues, and providing an interest-free health services to patients and that the member's activities must be in strict conformance with the law (Josen, 1990).

American Counseling Association – Ethical Approach

Counselors are often faced with situations that require sound ethical decision-making ability. Determining the appropriate course to take when faced with a difficult ethical dilemma can be a challenge. To assist American Counseling Association (ACA) members in meeting this challenge, the ACA Ethics Committee has developed A Practitioner's Guide to Ethical Decision Making. The intent of this document is to offer professional counselors a framework for sound ethical decision-making. The following will address both guiding principles that are globally valuable in ethical decision making, and a model that professionals can utilize as they address ethical questions in their work.

Moral Principles

Kitchener (1984) has identified the autonomy, non-maleficence, beneficence justice and fidelity as the most important aspects of moral principles that help clarify issues in a given situation. The autonomy allows an individual the freedom of choice and action, and that counselors are advised to encourage clients, when appropriate, to make their own decisions and to act on their own values so long as they have the capability. Non-maleficence is unintentional harm, and avoiding actions that propagates it. Beneficence is the contribution to the client's welfare; proactive and preventive of harm when possible. Justice indicates that all individuals should be treated the same. Whenever an individual is to be treated different out of the norm, the counselor should be able to provide a reasonable explanation for that choice. Finally, fidelity involves loyalty, faithfulness, and honoring commitments and that promotes trust between the counselor and clients. The counselor is advised to use the above as guidelines when exploring an ethical dilemma as doing so will help in working through the steps of an ethical decision making model, and to determine the moral principles in conflict (Forester & Davis, 1996).

International Ethical Guidelines

To broaden the understanding of international ethics, I tapped into the International code of ethics for the Occupational Health Professionals in addition to several others. According to Kogi (2002), the aim of occupational health practice is to protect and promote workers' health, to sustain and improve their working capacity and ability to contribute to the establishment and maintenance of a safe and healthy working environment for all, and further promote the adaptation of work to the capabilities of workers, taking into account their state of health. The field of occupational health is broad and covers the prevention of all impairments arising out of employment, work injuries and work-related disorders, including occupational diseases and all aspects relating to the interactions between work and health. In addition, Kogi (2002) mentioned that occupational health professionals should be involved, in the design and choice of health and safety equipment, procedures and safe work practices, and encourage workers' participation in this field and solicit feedback from their experience.

Kogi (2002) also indicated that there are several reasons why the International Commission on Occupational Health (ICOH) has adopted an International Code of Ethics for Occupational Health Professionals, as distinct from codes of ethics for all medical practitioners. One is the increased recognition of the complex and sometimes competing responsibilities of occupational health and safety professionals towards workers, employers, public, public health and labor authorities and other bodies such as social security and judicial authorities. Another reason is the increasing number of occupational health and safety professionals resulting from the compulsory or voluntary establishment of occupational health services. The Code applies to occupational health professionals and occupational health services regardless of whether they operate in a free market context subject to competition or within the framework of public sector health services.

Professional Ethical Conduct for the American Medical Informatics Associations

According to Thomas, Sage, Dillenberg, and Guillory (2002), it is moral to protect public health as it also protects the well being of communities and the country at large. Public health ethics ensures that the health of populations is not compromised and the abuse of power is evident in our societies (Thomas et. al, 2002).

The American Medical Informatics Association (AMIA) and other large professional societies have worked vehemently to promote a strong ethical framework for their membership. The AMIA manages clinicians, scientists, researchers, educators, students, and other informatics professionals and they rely on data to connect people, information, and technology. It manages more than 5,000 health care professionals in communities and the health care industry. It drives innovation for future management in biomedical research, clinical care, and public health. As the voice of the nation's top biomedical and health informatics professionals, AMIA members play a leading role in the following:

- Moving basic research findings from bench to bedside;
- Evaluating interventions across communities;
- Assessing the impact of health innovations on health policy; and
- Advancing the field of informatics, just to name a few.

Women and minority as research subjects

According to the National Institute od Health (NIH, 2001), it is their policy that that women and members of minority groups and their subpopulations must be included in all NIH-funded clinical research, unless a clear and compelling rationale and justification establishes to the satisfaction of the relevant Institute/Center Director that inclusion is inappropriate with respect to the health of the subjects or the purpose of the research. Further, cost is not an acceptable reason for exclusion except when the study would duplicate data from other sources.

Investigators, sponsors or ethical review committees should not exclude women of reproductive age from biomedical research. The potential for becoming pregnant during a study should not, in itself, be used as a reason for precluding or limiting participation. However, a thorough discussion of risks to the pregnant woman and to her fetus is a prerequisite for the woman's ability to make a rational decision to enroll in a clinical study. If participation in the research might be hazardous to a fetus or a woman if she becomes pregnant, the sponsors/investigators should guarantee the prospective subject a pregnancy test and access to effective contraceptive methods before the research commences. Where such access is not possible, for legal or religious reasons, investigators should not recruit for such possibly hazardous research women who might become pregnant (NIH, 2001). Women in most societies have been discriminated against with regard to their involvement in research. Women who are biologically capable of becoming pregnant have been customarily excluded from formal clinical trials of drugs, vaccines and medical devices owing to concern about undetermined risks to the fetus. Consequently, relatively little is known about the safety and efficacy of most drugs, vaccines or devices for such women, and this lack of knowledge can be dangerous (NIH, 2001).

Individual consent of women

In research involving women of reproductive age, whether pregnant or non-pregnant, only the informed consent of the woman herself is required for her participation. In no case should the permission of a spouse or partner replace the requirement of individual informed consent (Levin, 2008). If women wish to consult with their husbands or partners or seek voluntarily to obtain their permission before deciding to enroll in research, that is not only ethically permissible but in some contexts highly desirable. A strict requirement of authorization of spouse or partner, however, violates the substantive principle of respect for persons (Levine, 2008). A thorough discussion of risks to the pregnant woman and to her fetus is a prerequisite for the woman's ability to make a rational decision to enroll in a clinical study (Levine, 2008). For women who are not pregnant at the outset of a study but who might become pregnant while they are still subjects, the consent discussion should include information about the alternative of voluntarily withdrawing from the study and, where legally permissible, terminating the pregnancy (Levine 2008). Further, according to Levine (2008), if the pregnancy is not terminated, they should be guaranteed a medical follow-up. The investigator must establish secure safeguards of the confidentiality of subjects' research data. Subjects should be told the limits, legal or other, to the investigators' ability to safeguard confidentiality and the possible consequences of breaches of confidentiality

Confidentiality between Investigators and Subjects

According to Parker (2007), research relating to individuals and groups may involve the collection and storage of information that, if disclosed to third parties, could cause harm or distress. Investigators should arrange to protect the confidentiality of such information by, for example, omitting information that might lead to the identification of individual subjects, limiting access to the information, anonymity of data, or other means. During the process of obtaining informed consent the investigator should inform the prospective subjects about the precautions that will be taken to protect confidentiality.

Confidentiality between physician and patient

According to Parker (2007), *p*atients have the right to expect that their physicians and other health-care professionals will hold all information about them in strict confidence and disclose it only to those who need, or have a legal right to, the information, such as other attending physicians, nurses, or other health-care workers who perform tasks related to the diagnosis and treatment of patients. A treating physician should not disclose any identifying information about patients to an investigator unless each patient has given consent to such disclosure and unless an ethical review committee has approved such disclosure (Parker, 2007). Physicians and other health care professionals record the details of their observations and interventions in medical and other records. Epidemiological studies often make use of such records. For such studies it is usually impracticable to obtain the informed consent of each identifiable patient; an ethical review committee may waive the requirement for informed consent when this is consistent with the requirements of applicable law and provided that there are secure safeguards of confidentiality. In institutions in which records may be used for research purposes without the informed consent of patients, it is advisable to notify patients generally of such practices; notification is usually by means of a statement in patient-information brochures.

For research limited to patients' medical records, access must be approved or cleared by an ethical review committee and must be supervised by a person who is fully aware of the confidentiality requirements (Parker, 2007).

Ethical Issues in Community Healthcare

In a study about ethics in community research engagement, Smith and Blumenthal (2012) recalled the past history of minority exploitation in research such as the Tuskegee and Syphilis study's negative impact on minorities towards future community research and researchers about minority involvement. To determine whether minority communities are fairly treated in current researches, they explored the experiences and lessons learned from community health workers (CHWs) in a 10-year translation of an educational intervention in the research-to-practice-to-community continuum. They concluded that the central role played by CHWs enabled the community to gain some degree of control over research outcomes and implementation, indicating that the ethical principles of community-based participatory research (CBPR) has some positive impact. They further mentioned that Green and Mercer defined *community-based participatory research* (CBPR) as "a systematic inquiry, with the collaboration of those affected by the issue being studied, for purposes of education and taking action or effecting change." Community-based participatory research is effective and could reduce health disparities if partnerships in planning, implementation, evaluation, and dissemination between investigators and community members are collaboratively done.

Community health workers are now widely used in both research and public health practice involving minority groups indicating that research programs in minority communities are gaining grounds by winning the trust of the communities that had been lost for decades. Community members are being offered several benefits from the CHWs programs. Both the community members and the investigators continue to benefit from the program due to its fairness. It provides employment opportunities and skill developments for the community workers, and the investigators continue to gain community members' trust, and deploying CHWs enhances access to target populations and promotes research participation capacity for community development.

The investigators indicated that community leaders should be mindful of the following guidelines when engaging in a community research partnership with potential investigators:

- Ethical considerations should be a priority in all activity engagement
- Community commitment to the project by all stakeholders is a must
- CHWs must be empowered to bridge the research-to-practice gap

- Training is fundamental and should be strategically implemented
- Technical assistance is a requirement and should be available throughout the project

To conduct a community research, several factors are required to play a role, however, the ethical imperatives associated with it should be given the utmost priority. Community leaders must avoid exploitation at all cost during the course of any research and that researchers should not benefit at the expense of the community. Power and resources must be transferred to the community and power-sharing rates, if any, must be equitable distributed between the academia and the community involved. Finally, ensure that the project adheres to the principles of research ethics in the community.

In another community ethics research study pertaining to administrators' role in community health services, Aroskar (1998) found that patients receiving care from community health agencies such as home care present some special ethical challenges to clinicians and administrators as a result of the setting for care and other community factors. Further, administrators have less consideration of their ethical obligations and responsibilities compared to clinicians as they are in a position of making decisions and developing policies that impact the well being of both patients and employees. Healthcare administrators are found in the private, public, profit, non-profit sectors of the healthcare industry. Some are employed in managed care arrangements, integrated service networks, and community agencies. Aroskar further noted that healthcare administrator's daily work activity is filled with ethical decision making in healthcare entity, however, they turn to over use power and influence over clinicians and other managers, and that has negatively impact on the delivery of healthcare services in communities. Because ethics and power dissemination in work place is under the responsibilities of healthcare administrators in terms of decision-making and policy development, misuse of responsibilities has been reported and a guide for job responsibilities needs to be implemented and enforced. The Joint Commission for Accreditation of Healthcare Organizations (JCAHO) mandates include development of ethical standards for home care agencies who seek accreditation, and that administrators and clinicians must establish effective processes and mechanisms for responding to responsibilities as "moral agents" of the organization, yet, additional mechanisms for monitoring administrators for efficient and effective output needs to be put in place. This will go a long way to ensure favorable work place environment and there by enhancing quality care delivery to patients in the community (Aroskar, 1998).

Breadth Summary

I broadly presented a review of ethical theories on various issues in health organizations. Theories from John Rawls, Immanuel Kent regarding patients' informed consent and other major ethical dilemmas have been synthesized in this breadth component.

Reviewing these ethical frameworks have brought to light the notion that ethical principles are unique to individual organizations although there may be some common grounds for various concepts. In the next two components, however, I strengthen the connection between the aforementioned theoretical frameworks about ethics to specific ethical problems in health care. Further, I provided a review of current research findings regarding ethical diversity issues in healthcare, and this will serve as a reference to healthcare providers.

Depth

Health Services and Ethical Diversity Decision-Making

I N ORDER TO acquire a broader and a better understanding of ethical delivery in health services, I reviewed several articles from current research ranging from assessment of harm potential and factors influencing participation decisions to the improvement of patients' health concerns. The ethical concepts reviewed in the breadth components provided a broad understanding of health care ethics. In the depth component, however, a review of current research findings narrowed the focus. The following are the bibliography of contemporary research outcome on ethical issues in health services delivery:

Annotated Bibliography

Blanco and Suresh (2005) found that physicians, nurses, and nurse practitioners underestimated survival rates and overestimated long-term disability rates for intensive premature infants. After education, their estimates of survival and long-term disability rates for these infants improved significantly. They noted that more accurate estimates of survival rates of premature infants by physicians' and nurses' theoretical decision-making appropriateness are required.

A study by Brody and Scherer (2006) on family and physician influence on asthma research participation decisions for adolescents indicated that adolescents were less willing to cede decision making authority to parents than parents anticipated. Parents and adolescents acknowledged a greater openness to influence from physicians than from family for above minimal risk studies, and that Parents were more willing to consider opinions from male adolescents.

To determine the influence of ethical safeguards on research participation, Hammond and Lewis (2004) examined 60 people with schizophrenia, and 69 psychiatrists rated the protectiveness and influence on patients' willingness to participate in research of five safeguards: informed consent, alternative decision makers, institutional review boards, data safety monitoring boards, and confidentiality measures. The investigators found that ethical commitment to research volunteers is expressed in safeguards. These efforts appear to be viewed positively by key stakeholders and may influence research participation decision-making.

Janvier and Barrington (2005) conducted a study to determine the adequacy of records of parental counseling in mothers with threatened preterm delivery before 27 weeks gestation, whether interventions performed at birth were consistent with recorded antenatal decisions, and whether the extent of resuscitation affected the occurrence of serious short-term morbidity. The conclusion was that records of antenatal consultations were often lacking important information. Variations in physician documentation practices are substantial and affect the care offered to infants at the threshold of viability.

To evaluate systematic ethics reflection in groups in community health that includes nursing homes and residencies from the perspectives of employees participating in the groups, the group facilitators, and the service managers, Lillemoen and Pedersen (2015) applied a mixed-methods design with qualitative focus group interviews, observations and written reports at two nursing homes, two home care districts and a residence for people with learning disabilities. The researchers concluded that ethics reflection groups focusing on ethical challenges from the participants' daily work were found to be significant for improved practice, collegial support and cooperation, personal and professional development among staff, facilitators and managers. Further, resources needed to succeed were managerial support, and anchoring ethics sessions in the routine of daily work.

Scherer and Annett (2006) studied the considerable ethical and legal ambiguity surrounding the role of adolescents in the decision-making process for research participation. They examined parent and adolescent perceptions of decision-making authority and sources of influence on adolescent research participation decisions, and examined whether perceptions of influence differed based on adolescent gender and level of research risk. The researchers concluded that adolescents desire responsibility for research participation decisions, though parents may not share these views.

Scott and Kalawish (2004) conducted an electronic and manual literature search for all English-language articles examining the decision-making capacity of elderly persons with dementia or cognitive impairment, reviewing articles in relation to key areas of methodological, clinical, and policy importance. The researchers found that although, incapacity is common, many persons with dementia are capable of making their own medical and research decisions. Furthermore, in early stages of dementia, interventions may improve decisional abilities, and that simple cognitive screenings may be useful by identifying persons in need of more intensive evaluations.

Sharman and Meert (2005) studied the decisions to forgo life support from critically ill children that are commonly faced by parents and physicians, and they identified and described parents' self-reported influences on decisions to forgo life support from their children. The results indicated that previous experience with death and end-of-life decision making for others, their personal observations of their child's suffering, their perceptions of their child's will to survive, their need to protect and advocate for their child, and the family's financial resources and concerns regarding life-long care influence parents decision to forgo life support. The researchers concluded that inclusion of factors like past experiences with end-of-life decisions, and their anticipated emotional adjustments and future resources into discussions is important to parents and may facilitate decisions regarding the limitation or withdrawal of life support.

Clinical Issues in Informed Consent

Appelbaum (2007) stated that law and medical ethics require physicians to obtain the informed consent of their patients before initiating treatment, and that when patients lack the competence to make a decision about treatment, substitute decision makers must be sought. Hence, the determination of whether patients are competent is critical in striking a proper balance between respecting the autonomy of patients who are capable of making informed decisions and protecting those with cognitive impairment. Valid informed consent is premised on the disclosure of appropriate information to a competent patient who is permitted to make a voluntary choice.

Dunn and Roberts (2005) reviewed several publications on the following six topics: professional integrity and sources of potential bias; scientific designs; protocol safeguards; influences on research participation decisions and perceptions of risk; informed consent decision-making capacity, appreciation and the therapeutic misconception, and voluntarism; and informed consent-intervention studies, and found that little empirical work has been conducted on investigator training and potential conflicts of interest. Numerous concepts and controversial issues related to the study designs with the initial focal point being the ethics of placebo-controlled studies have been tackled and that only a handful of reports, however, have empirically addressed these topics, therefore, empirical studies focusing on a variety of ethically relevant domains in schizophrenia research are needed.

In a study to understand mothers' and counselors' perceptions of their roles in decision-making about resuscitation of extremely premature infants at delivery and to assess mothers' and counselors' satisfaction with the counseling and decision-making process, Keenan and Doron (2005) interviewed mothers who delivered an infant between 22 and 27 completed weeks of gestation and their self-identified counselors using a structured interview format. They concluded that decision-making process in this study conforms most closely to a model of informed assent, and that the mothers have been satisfied with this type of counseling because they felt informed and included in the decision-making process. Further, physicians and nurses need to elicit mothers' preferences to incorporate them into the treatment plan, as counseling about the resuscitation of extremely premature infants at delivery is considered directive by mothers even when it is not intended to be directive.

Nijhawan et al. (2013) found that for a drug to get approved and enter into the market it has to prove its safety and efficacy in clinical trials, and that is a research related activities using human being as subjects. Therefore, no one has the right to infract fundamental rights of another person for the sake of fulfilling their own purpose, and that brought about the use of informed consent tool. Informed consent is not only required for clinical trials but an essential prerequisite before enrolling each and every participant in any type of research involving human subjects including; diagnostic, therapeutic, interventional, bioequivalence, social and behavioral studies and for all research conducted domestically or abroad. Obtaining consent involves informing the subject about his or her rights, the purpose of the study, the procedures to be undergone, the potential risks and/or benefits of participation and alternative treatments available if any.

Davis, Berkel and Holcombe (1998) indicated that to ensure that patients fully understand factors related to their care, the Food and Drug Administration (FDA) requires that consent documents contain detailed information regarding eight basic elements of informed consent but little attention has been given to how well patients comprehend these elements, although, health care providers have an ethical and legal responsibility to ensure that patients understand their participation in research. Their study findings raise serious questions regarding the adequacy of the design of written informed consent documents for the substantial proportion of Americans with low-to-marginal literacy skills, and that a high level of reading skill and comprehension is required to understand and complete most consent forms that are required for participation in clinical research studies. The study further indicated that simplifying informed consent material alone makes the forms more appealing and easier to read but will not improve comprehension, and that research is needed to redesign informed consent contents to increase comprehension, especially for participants with inadequate or marginal reading skills.

Annotated Bibliography Summary

It is obvious that ethics in health care delivery yearns for constant monitoring and upgrading as the review indicated. This upgrading involves educating health care providers, patients, and communities to strive for the best in health care delivery. Some of the ethical issues that require immediate attention include the infant survival rate as a result of physicians' estimation influenced by inappropriate record keeping and documentation practices, inadequate intervention services to incapacitated patients so they can make decisions on their own when the need arises, appropriate parental guidelines to parental decisions to forgo life support of their loved ones at the point of death. Informed consent publications should be free from ambiguity and be more clarity to patients and research subjects. Further, appropriate ethical protocol should be adhered to in obtaining informed consent from patients prior to initiating treatment.

Literature Review

In addition to the annotated bibliography articles reviewed above, the following articles further provided clarification on the need for further research for a better understanding of ethical decision-making on the part of clinicians and health care providers in general. According to Kim and Karlawish (2002), the importance of research is enormous in that it improves the health of a nation's population as a whole by fostering a better health system. Additionally, emphasis is placed on the devastation of withholding information from patients and research participant. Researchers and health professionals are seriously informed that patients and research participants should be well informed about the risk involve in research participation and or health related procedure.

Global Health and Primary Care Research

Countries around the world are advised to strive for effective and efficient healthcare system with their limited resources. A strong primary health care system is important to have and should constantly be improved and that will eventually improve the health of the population. It is important to invest in research in primary care for that will go a long way to develop better healthcare systems and health policies. (Kim & Carlawish, 2002). Commitment by international organizations within and outside of primary care is required to fund research to enhance organizations involved. Kim and Carlawish suggested the following with regards to improving the global primary care research enterprise for the benefit of the world's population:

- Ensure the validity of research by withholding certain information in the consent process and must receive an approval of the ethical review committee. Withholding information from research subjects ensures unbiased outcome.
- Researchers must bear in mind that research subjects must not be intimidated for doing so invalidates informed consent. Research subjects must be informed that their decision to withdraw from a research program can be done at any time without penalty.

Investigators should be completely objective in discussing the details of the experimental intervention, the pain and discomfort that it may entail, and known risks and possible hazards. In complex research projects it may be neither feasible nor desirable to inform prospective participants fully about every possible risk. They must, however, be informed of all risks that a 'reasonable person' would consider material to making a decision about whether to participate, including risks to a spouse or partner Van Hoose, 1980). Before proceeding without prior informed consent, the investigator must make reasonable efforts to locate an individual who has the authority to give permission on behalf of an incapacitated patient. If such a person can be located and refuses to give permission, the patient may not be enrolled as a subject. The researcher and the ethical review committee should agree to a maximum time of involvement of an individual without obtaining either the individual's informed consent or authorization according to the applicable legal system if the person is not able to give consent. If by that time the researcher has not obtained either consent or permission – owing either to a failure to contact a representative or to a refusal of either the patient or the person or body authorized to give permission – the participation of the patient as a subject must be discontinued. The patient or the person or body providing authorization should be offered an opportunity to forbid the use of data derived from participation of the patient as a subject without consent or permission (Parker, 2007).

Informed Consent and Malpractice Lawsuits

The following outcome of court proceedings has pointed out some of the repercussions of neglecting informed consent application in healthcare delivery. Healthcare entities will see a tremendous reduction in malpractice related costs, and maintain if not increase their competitive edge if the principles of informed consent are adhered to. The court trials outcome about informed consent negligence by healthcare providers will create awareness about the serious consequences involved.

Informed Consent Negligence and Outcome of Court Proceedings

In the Court case of Schloendorff v. Society of New York Hospital, 105 N.E. 92, 1914, published in a basic case in the bioethics literature, Justice Cardozo quoted that "Every human being of adult years and sound mind has a right to determine what shall be done with his own body...." This statement upholds the concept of informed consent in the healthcare delivery, and that risk management in the industry must be promoted vehemently.

Canterbury v. Spence, 464 F.2d 772, 1972

In this well-known case, the Court argued for a *reasonable patient* standard of informed consent. In doing so, the Court presented arguments against a *physician-oriented* standard. The case involves a boy who suffered paralysis after back surgery. He claimed that he was not adequately warned about such risks. Although recognizing that a majority of courts look to customary practices of physicians in dealing with informed consent, and that deviation from such standards may be a cause for legal action, the *Canterbury* Court held that informed consent depends on the individual circumstances of a patient. "In our view, the patient's right of self-decision shapes the boundaries of the duty to reveal." A physician may be held liable whenever the informed consent process unreasonably neglects what the physician should know to be the patient's informational needs.

Cobbs v. Grant, 8 Cal. 3d 229, 1972

The patient alleged that the surgeon violated his duty to obtain informed consent to surgery by failing to inform the patient of its inherent risks, many of which materialized. The Court concluded that the informational needs of the patient are paramount in obtaining informed consent. Thus, standards that emphasize the conduct of the physician are inappropriate. The legal rule that emerges from *Cobbs* is that a physician is under duty to disclose to a patient those risks and benefits of the proposed treatment, alternative treatments, and of no treatment, which a *hypothetical reasonable patient* would consider material.

Backlund v. University of Washington, 975 P.2d 950, 1999

The Court considered whether failure to provide adequate information was a cause of the patient's injury. This does not necessarily involve negligence. Treatment may be applied in a non-negligent way and still result in an adverse outcome. If a person were informed of the risk, and rejected treatment, the harmful event would not have resulted. In this way, lack of informed consent may have caused the injury. The Court must determine whether the patient would have rejected treatment had information about risk been provided. In this case, the Court opted for an *objective* as opposed to a *subjective* standard of causation. A subjective standard involves what the given patient would have done. This is difficult to determine and may lead to false statements by the plaintiff. An objective standard was adopted (as it is in many jurisdictions) involving what a reasonable patient would have done in the patient's situation.

Gouse v. Cassell, 615 A.2d 331, 1992

The Pennsylvania Supreme Court explicitly rejected the idea the patient must show that he or she would not have had the treatment had informed consent been thorough: "We hold that a physician or surgeon who fails to advise a patient of material facts, risks, complications and alternatives to surgery which a reasonable man in the patient's position would have considered significant in deciding whether to have the operation is liable for damages which ensue, and the patient need not prove that a causal relationship exists between the physician's or surgeon's failure to disclose information and the patient's consent to undergo surgery."

Blanchard v. Kellum, 975 S.W.2d 522, 1998

A woman had all 32 of her teeth extracted during a dental visit. She claimed that she never authorized that all of her teeth would be extracted in a single visit. The extractions caused her to be hospitalized. The issue at stake is whether this case should be handled as negligence or as battery. The Court observed: "We believe that there is a distinction between: (1) cases in which a doctor performs an unauthorized procedure; and (2) cases in which the procedure is authorized but the patient claims that the doctor failed to inform the patient of any or all the risks inherent in the procedure." Furthermore, "The primary consideration in a medical battery case is simply whether the patient knew of and authorized a procedure. This determination does not require the testimony of an expert witness."

Lugenbuhl v. Dowling, 701 So.2d 447, 1986

What if a physician does not follow the treatment agreed to even though no medical damages were involved? In this case a patient was awarded damages due to a battery-like claim. A procedure that the patient explicitly rejected was used in a hernia surgery. The Court concluded that there was no evidence that the technique used caused the patient's further complications. In effect, damages stemmed from depriving the patient of the "opportunity of self-determination...." This seems to be a battery claim, so the damage award did not depend on whether the treatment itself caused further medical problems.

Since this case did not involve malpractice, the Court significantly reduced the monetary award.

Ashcraft v. King, 278 Cal.Rptr. 900, 1991

Surgery was performed after it was explicitly stated that only blood donated by family members should be used in any transfusion. The physician used blood from the hospital's general supply. HIV resulted. Although the operation was consented to, the plaintiff claimed battery. The Court declared: "In an action for civil battery the element of intent is satisfied if the evidence shows defendant acted with a 'willful disregard' of the plaintiff's rights." In this case, the Court found that such willful disregard occurred.

Mohr v. Williams, 104 N.W. 12, 1905

During surgery, and after consent to operate on a patient's right ear, a surgeon determined that the patient's left ear also required treatment. The treatment was successful, yet performed without consent. The Court rejected the contention that lack of damage meant that a battery did not take place: "The last contention of defendant is that the act complained of did not amount to an assault and battery. This is based upon the theory that, as plaintiff's left ear was in fact diseased, in a condition dangerous and threatening to her health, the operation was necessary, and, having been skillfully performed at a time when plaintiff had requested a like operation on the other ear, the charge of assault and battery cannot be sustained. We are unable to reach that conclusion, though the contention is not without merit. It would seem to follow from what has been said on the other features of the case that the act of defendant amounted at least to a technical assault and battery."

Duncan v. Scottsdale Med. Imaging, Ltd., 70 P.3d 435, 2003

A patient requested to know the kind of drug a nurse was administering. She was told that it was fentanyl, a synthetic drug similar to morphine. The patient rejected that drug, and asked the nurse to call the physician. The nurse subsequently told the patient that the drug was changed to morphine. The patient then consented. However, the drug was not changed, and instead the nurse administered fentanyl. A battery claim was permitted. The Court followed *Cobbs v. Grant* in claiming that there was a distinction between "lack of informed consent" and "lack of consent." "The battery theory should be reserved for those circumstances when a doctor performs an operation to which the patient has not consented." The Court lamented that this distinction is sometimes blurred.

Shine v. Vagas, 429 Mass. 456, 1999

The Supreme Court of Massachusetts clearly rejects the notion that in emergency situations competent persons may be treated despite their refusal of treatment. The Court considered the emergency conditions under which consent is not required based on the doctrine of presumed consent: "If, and only if, the patient is unconscious or otherwise incapable of giving consent, and either time or circumstances do not permit the physician to obtain the consent of a family member, may the physician presume that the patient, if competent, would consent to life-saving medical treatment."

Wells v. Van Nort, 100 Ohio St. 101, 1919

A woman consented to surgery to remove her appendix, but during the surgery her fallopian tubes were removed because they were found to be diseased. The Court held that the woman did not consent to have her fallopian tubes removed and that this was not an emergency situation.

Truman v. Thomas, 165 Cal. Rptr. 308, 1980

In some jurisdictions, a patient must be informed about the risks of non-treatment. In this case, a physician, Dr. Thomas, did not adequately explain to the patient, Mrs. Truman, the risks involved with failure to undergo a pap smear. When a patient refuses treatment, the Court asserted that there is a duty to advise about material risks that a reasonable person would want to know. The Court based this partly on a fiduciary obligation, due to the knowledge of the physician and the lack of knowledge of the patient. "It must be remembered that Dr. Thomas was not engaged in an arms-length transaction with Mrs. Truman. Clearly ... he was obligated to provide her with all the information material to her decision [not to have a pap smear]."

Moore v. Regents of the University of California, 271 Cal.Rptr. 146, 1991

This case involved a patient with leukemia. His physician had an economic research interest in the patient's blood and bone marrow aspirate. This interest was not disclosed. The Court concluded: "A physician who adds his own research interests to this balance may be tempted to order a scientifically useful procedure or test that offers marginal, or no, benefits to the patient. The possibility that an interest extraneous to the patient's health has affected the physician's judgment is something that a reasonable patient would want to know in deciding whether to consent to a proposed course of treatment. It is material to the patient's decision and, thus, a prerequisite to informed consent."

Neade v. Portes, 237 Ill.Dec. 788, 1999

In contemporary health care arrangements, health care professionals may have a financial incentive to avoid diagnostic and treatment procedures. In this case, with little precedent, the Court struggled with the issue of whether there is a fiduciary duty to inform a patient about financial incentives. The Court supported disclosure of such financial interests, and cited the Current Opinions of the Council on Ethical and Judicial Affairs of the American Medical Association (1996-1997 edition). Section 8.132 of Current Opinions provided that patients should not be denied access to appropriate medical care based on personal financial gain or loss.

Healthcare providers/physicians have a fiduciary duty to provide appropriate care in addition to adhering to the principle of informed consent. Considering the enormous court cases resulting from informed consent application and other medical malpractices, it is advisable for the healthcare providers to constantly utilize or take advantage of healthcare workshops at their workplaces to serve as a guide to appropriate care. Healthcare entities have gone out of business as a result of medical malpractices that could have been prevented had the education of healthcare providers been taken seriously. Novel healthcare providers are advised to consider patients as their primary focus and do whatever it takes to accommodate them appropriately.

Depth Summary

The Depth section integrated ethical concepts of classical theories by John Rawls and Immanuel Kant with the findings of current ethical research. Some of the topics of discussion in the literature review included the research participation decision-making, appropriate delivery of information participants and the need for global health and primary care research. Furthermore, some of the current research reviews were on the following topics: Parental involvement in treatment decisions regarding their critically ill child and ethical issues in neonatal intensive care and physician's practices and informed consents and malpractice lawsuits facing healthcare providers. The findings of these studies may provide opportunity for physicians to evaluate their practices critically. The application section utilizes the findings of current research in the depth section, especially in the area of ethical implication in clinical practice and research approach and the theoretical concepts reviewed in the breadth component to provide guidelines to health care professionals. Adhering to these guidelines may alleviate some of the ethical dilemmas in health care.

Application

Ethical Theories Applied in Health Services

Introduction

T O ACCOMPLISH THE objective of recommending an effective approach to ethical dilemma, I considered a further review of additional code of ethics below. I have pointed out in the breadth and depth components about the importance of ethical diversity in healthcare and the guidelines to an ethical decision in health care, and healthcare provider's awareness of these guidelines. In the application component, however, I further reviewed ethical models from both national and international levels. I reviewed the Australian code of ethics into details to provide an insight on international ethics. Traditional and religious ethical concerns have also been reviewed. Additionally, I culminated with the broad ethical concepts reviewed in the depth component and the current research analyzed in the depth component and provided some vital recommendations and guidelines that could be utilized by the health care industry.

AMA Code of Ethics

Medical doctors undergo a specialized training, as such, they have a professional responsibility to maintain and improve the health of patients in whatever situation in terms of illness and welfare situations. Healthcare providers must make every effort to improve patients' health. For several years, medical doctors have gone through ethical principles training and development primarily to guide them in delivering appropriate care to patients. The codes of ethics reviewed in the depth component guide doctors to promote the health and well being of their patients and prohibit doctors from unnecessary mistakes that will harm the patient. I learned a great deal by reviewing the AMA code of ethics and comparing it to other health organizations in the United States. This move has helped in recommending an adequate code of ethics for organizations to adopt. It is without a doubt that implementing changes in medical management brings with it new and challenging ethical problems, Professional Conduct Committees have to strategize in their implementations new guidelines. Implementation should be spread out for a period of time instead of bulky methodology. Health care professionals from various healthcare organizations have the responsibilities of adhering to the following code of ethics in addition to others according to (AMA) (1996).

The Doctor and the Patient – Standard of Care

1. Evaluate your patient completely and thoroughly, maintain accurate contemporaneous clinical records.

2. Ensure that doctors and other health professionals who assist in the care of your patient are properly qualified and fully competent to carry out the care.

3. Do not exploit your patient for sexual, emotional or financial reasons.

4. Treat your patient with compassion and respect for human dignity.

5. Respect your patient's right to choose doctors freely, to accept or reject advice and to make their own educated decisions about treatment or procedures.

6. In general, keep in confidence information derived from your patient, or from a colleague regarding your patient, and divulge it only with the patient's permission, except when a court demands.

7. When a personal moral judgment or religious conscience alone prevents the recommendation of some form of therapy, inform your patient so that they may seek alternative care.

8. Obtain prior consent of all research subjects or their agents, but only after explaining the purpose of the clinical research and any reasonably foreseen health hazards.

9. Do not allow a refusal to participate at any stages interfere with the doctor-patient relationship or appropriate treatment and care.

Caring For The Dying Patient

The process of advance care planning is widely recognized as a way to support patient self-determination, facilitate decision-making, and promote better care at the end of life (American Medical association (AMA, 2016). Not only applied to terminally ill patients or those with chronic medical conditions, advance care planning is valuable for everyone, regardless of age or current health status. It is important to give patients the opportunity to identify who they would want to make decisions on their behalf should in case they do not have decision-making capacity as a result of terminal illness. Patients should be encouraged to share their views with families or other intimates and to name a surrogate decision maker to help ensure that patients' own values, goals, and preferences are applied when needed. Physicians must be aware that patients and families approach decision-making in many different ways that are influenced by culture, faith traditions, and life experience, and should be sensitive to each patient's individual situations and preferences when planning for end of life care.

Physicians must remember to incorporate notes from the advance care planning discussion into the medical record. Further, patient values, preferences for treatment, and designation of surrogate decision maker should be included in the notes to be used as guidance when they are unable to express their own decisions. Patients advance directive document or written designation of proxy should also be included in the medical records and advise patients to give a copy to their surrogates and relatives they are comfortable with to help ensure its availability when needed. When patients without decision-making capacity, advance directives and no surrogate available, and the attending physician is to make treatment decisions on their behalf, it is advisable to seek assistance from ethics committee or other appropriate resource to ascertain the patients' best interest.

According to the American Medical Association (AMA), decisions to withhold or withdraw life-sustaining interventions can be ethically and emotionally challenging to all involved. However, a patient who has a decision-making capacity appropriate to the decision at hand has the right to decline any medical intervention or ask that an intervention be stopped regardless of whether or not the patient is terminally ill. When a patient lacks appropriate capacity, the patient's surrogate may decline an intervention or ask that an intervention be stopped in keeping with ethics guidance for surrogate decision-making.

It is understandable, though tragic, that some patients in extreme duress, such as those suffering from a terminal, painful, debilitating illness, may come to decide that death is preferable to life. However, permitting physicians to engage in assisted suicide would ultimately cause more harm than good. Physician-assisted suicide is fundamentally incompatible with the physician's role as healer, would be difficult or impossible to control, and would pose serious societal risks. Physicians must aggressively respond to the needs of patients at the end of life in the following manner as specified by AMA Principles of Medical Ethics instead of engaging in assisted suicide. Physicians:

- Should not abandon a patient once it is determined that cure is impossible.
- Must respect patients' autonomy.
- Must provide good communication and emotional support.
- Must provide appropriate comfort care and adequate pain control.

Further Advice About The Dying Patient

Always bear in mind the obligation of persevering but, allow death to occur with dignity and comfort, is deemed to be inevitable and where curative appears to be futile.

Transplantation

1. Accept that when brain death has occurred (defined as the irreversible cessation of all functioning of the body including

brain stein, unless otherwise defined cellular life in the body may be supported if some parts of the body may be used to prolong life or to improve other people.

2. Recognize the responsibility to provide to the relatives a full disclosure of the intent to transplant organs, the purpose of the procedure and, in the case of the donor, the risks of the procedure.

3. Ensure that the determination of the time of death of any donor patient is made by doctors who are in no way concerned with the transplant procedure or associated with the proposed recipient in a way that may exert any influence upon existence made.

International Instruments and Guidelines

The first international instrument on the ethics of medical research, the Nuremberg Code, was promulgated in 1947 as a consequence of the trial of physicians (The Doctors' Trial) who had conducted atrocious experiments on un-consenting prisoners and detainees during the Second World War. The Code, designed to protect the integrity of the research subject, set out conditions for the ethical conduct of research involving human subjects, emphasizing their voluntary consent to research. The Universal Declaration of Human Rights was adopted by the General Assembly of the United Nations in 1948. To give the Declaration legal as well as moral force, the General Assembly adopted in 1966 the International Covenant on Civil and Political Rights. Article 7 of the Covenant states that "No one shall be subjected to torture or to cruel, inhuman or degrading treatment or punishment. In particular, no one shall be subjected without his free consent to medical or scientific experimentation". It is through this statement that society expresses the fundamental human value that is held to govern all research involving human subjects – the protection of the rights and welfare of all human subjects of scientific experimentation (Kogi, 2002). Since the publication of the Council for International Organizations of Medical Sciences 1993 Guidelines, several international organizations have issued ethical guidance on clinical trials. This has included, from the World Health Organization, in 1995, Guidelines for Good Clinical Practice for Trials on Pharmaceutical Products; and from the International Conference on Harmonization of Technical Requirements for Registration of Pharmaceuticals for Human Use (ICH), in 1996, Guideline on Good Clinical Practice, designed to ensure that data generated from clinical trials are mutually acceptable to regulatory authorities in the European Union, Japan and the United States of America. The Joint United Nations Program on HIV/AIDS published in 2000 the UNAIDS Guidance Document on Ethical Considerations in HIV Preventive Vaccine Research (Kogi, 2002).

In 2001 the Council of Ministers of the European Union adopted a Directive on clinical trials, which will be binding in law in the countries of the Union from 2004. The Council of Europe, with more than 40 member States, is developing a Protocol on Biomedical

Research, which will be an additional protocol to the Council's 1997 Convention on Human Rights and Biomedicine. This is a proof of the vital need of ethical guidelines both at the national and international level (Kogi, 2002).

Informed Consent's General Considerations

Informed consent is a decision to participate in research taken by a competent individual who has received and understood the necessary information from the researchers and has decided to participate without coercion. It is based on the principle that competent individuals are entitled to choose freely whether to participate in research. The individual freedom of choice is totally protected and respected. Research proposal should always be guided by the ethical review boards be it colleges or other organizations to make sure the informed consent aspect is appropriately addressed and no harm is afflicted to human subjects as they include young children, adults with severe mental or behavioral disorders, and persons who are unfamiliar with medical concepts and technology (Wirshing and Liberman, 1998).

It is advisable for Health care providers and researchers to be familiar with the following guidelines when making ethical decisions: During process, investigators should elicit information from the subjects with respect and dignity. Prospective subjects' questions and concerns throughout the research must be addressed as they arise and the subjects can withdraw from the study at any time without intimidation or penalty. Each individual must be allowed to consult family members and love ones to help them in reaching appropriate decision. In terms Language appropriate, investigators must be mindful of conveying information whether orally or written using appropriate language at the subjects' level of understanding and intelligence. The prospective subjects must understand the communication contents and willingly consent to the request. Again, the investigator must give full opportunity to subjects to ask questions and address them honestly, promptly and completely. The program must not go on if subjects are not in full understanding of what the program entails. Furthermore, subjects may imply consent by expressing consent orally or in a signature form. However, the ethical review committee may approve waiver of the requirement of a signed consent form if the research carries no more than minimal risk. Such waivers may also be approved when existence of a signed consent form would be an unjustified threat to the subject's confidentiality. Wordings on the informed consent form should be free from ambiguity, and that investigators are responsible for providing documentation or proof of consent prior to implementations. In some cultures an investigator may enter a community to conduct research or approach prospective subjects for their individual consent only after obtaining permission from a community leader, a council of elders, or another designated authority. Such customs must be respected. However the permission of a community leader or other authority must not replace the subject's informed consent. Subjects' cultural backgrounds or values should be taken into consideration when coming up with a written document, be it inform-consent or a survey questionnaire. Investigators must ensure that jargons are free from vague interpretations and are well understood (Appelbaum, 2007).

Recommended Guidelines for Obtaining Informed Consent

According to Rao (2008), before requesting an individual's consent in research participation, investigators must provide the following information in the languages or another form of communication understandable to the subjects. These guidelines will alleviate most of the problems that may arise in ethical decision-making by professionals during selection or participants for the research study.

1. A tangible reason for subjects' invitation and participation must be voluntary.
2. That the individual is free to refuse to participate and will be free to withdraw from the research at any time without penalty.
3. The duration of the research has to be clearly stated so as to allow participants leverage of making appropriate decision.
4. Providing money or other forms gifts for participation has to be clearly stated.
5. The findings from the study have to be shared with participants in addition to the entire community if applicable.

6. Subjects or participants should have the right of access to their data if demanded, even if these data lack immediate clinical application (unless the ethical review committee has approved temporary or permanent non-disclosure of data, in which case the subject should be informed of in advance.

7. Any foreseeable risks, pain or discomfort, or inconvenience to the individual (or others) associated with participation in the research, including risks to the health or well-being of a subject's spouse or partner must be thoroughly evaluated and discussed with participants.

8. Provisions should be made to ensure respect for the privacy of subjects and for the confidentiality of records in which subjects are identified.

9. There should a policy in place for the use of test results and other information involved, or sharing data with insurance companies or employers, and that has to be consented by the subjects.

10. The planning that biological specimens collected in the research will be destroyed at its conclusion, and, if not, details about their storage (where, how, for how long, and final disposition) and possible future use, and that subjects have the right to decide about such future use, to refuse storage, and to have the material destroyed.

11. Must be specified whether a treatment will be provided free of charge for specified types of research-related injury or for

complications associated with the research. The nature and the duration of such care and the organization responsible for the care should be provided. Further, a plan should be in place to compensate the family of participants for disability or death resulting from the research.

12. Finally, an ethical review committee should approve or clear the research protocol to proceed.

Cultural and Religious Competence in Healthcare

Studies indicate that a minimum level of cultural awareness is beneficial to healthcare services delivery by effectively addressing the needs of diverse patients. As such, healthcare providers are encouraged to broaden their knowledge about various cultures and religions. Padela Indicated that because societies are becoming increasingly multicultural with a plurality of value systems that may come into conflict with one another, cultural and religious differences could lead to bioethical conflicts. In order to bridge these differences each party must understand the ethical constructs that uphold each belief. Thus, not only are cultural and religious competences necessary, but also competence in dealing with divergent ethical codes. The field of medical ethics is a growing one, and healthcare providers are urged to realize and adjust to the trend.

In this chapter, I simplified and highlighted some key teachings in religious medical ethics and explored their applications. Hopefully, the insights gained will aid healthcare providers to better understand their patients who come form various religious backgrounds and to deliver appropriate care. The following are some fundamental information about medical ethics in various religions that healthcare providers should be aware of.

Medical Ethics in Islam

Islamic Medical Ethics calls for universal ideals within the framework of Islamic discourse. One important aspect within the Islamic Medical Ethics discourse is Adab literature. The literature codifies professional ethics and personal morality for both physicians and patients, and the reason for this dual emphasis is that the first is a belief in the ethical responsibility of the physician having two dimensions: one being the care and compassion that must embody the doctor's behavior towards the patient, and the physician must be righteous for his treatment to be efficacious, and that according Islamic teaching, ethical health is part of general health, and that physicians positive character plays a role on patients speedy recovery.

Medical ethics literatures call for a renewed interest in describing the varied ethical constructs of specific populations in providing an overview of Islamic Medical Ethics. The introduction of bioethical issues such as abortion, gender relations within the patient-doctor relationship, end-of-life care and euthanasia has opened an avenue of exploration and enlightenment. As the diversity in medicine grows, today's practicing clinician, particularly in the United States and Europe, encounters patients from a wide spectrum of socioeconomic and cultural backgrounds. As such, the cultural background of both the clinician and the patient influences the meanings attached to clinical encounter and interventions and that require cultural competence increased focus in medical education programs in order to provide effective clinical care to patients from diverse backgrounds.

Medical Ethics is concerned with moral principles as they relate to biomedical science in clinical underpinnings. Ethics relative to the Western ideology has developed into a philosophical science and has moved away from a Christian conception of good and evil to draw more upon human reason and experience in judging between right and wrong actions as pointed out in Emmanuel Kant's Metaphysics of Morals. This development from religious ethics to philosophical ethics is not in line with Islamic intellectual discourse. To define Islamic Medical Ethics appropriately, it is important to note that Islamic ethics as a cohesive discipline does not exist. Materials on ethics are located throughout the Islamic sciences of Fiqh (jurisprudential understanding), Tafsir (Quoranic exogenesis), and Kalam (scholastic theology). Similarly, ethicists in the Muslim world may refer to the Shari'ah (Islamic Law) when debating abortion, euthanasia (Intentionally ending one's life to relieve pain), end-of-life care, and other biomedical issues. Hence Islamic Medical Ethics is tied to Islamic Law, as Islamic Law not only legislates but also assigns moral values according to research.

Islamic Medical Ethics and the Islamic Law (Shari'ah)

The two important Arabic concepts that pose a challenge in their comprehension relative to Islamic Medical Ethics are the Fiqh and Shari'ah. Fiqh shapes Islamic Law and guide the ethics in healthcare delivery pertaining to Islam. It is the understanding of Islamic religious values or rulings or the sources of law statutes. Further, It represents the formulated legal rulings on a subject matter and/or the moral value assigned to a particular action. Shari'ah however, is the source of all life according to Islamic belief, and it determines the correct course of action. The Shari'ah, not only separates actions into required and forbidden, but also the intermediate categories of recommended, discouraged and permitted. Consequently, Islamic 'law' is both a legal and ethical system and any discussion of Islamic Medical Ethics must incorporate the Shari'ah.

The following issues regarding Islamic medical ethics were addressed at the First International Conference on Islamic Medicine in 1981 to synthesize a code of medical ethics from the Islamic perspective. Physicians' loyalty and faith were addressed; they must believe and follow measures geared towards patients' concern while adhering to the principles of Islamic ethics. Physicians and other healthcare providers must adhere to specific guidelines in addressing the Islamic patients' healthcare concern, and since Islam emphasizes its basic assumption as faith in Allah in addition to other pillars of Islam and morality, each individual as well as society should approach

Allah to the best of their abilities. This approach to Allah is made through the Shar'iah, and collectively, state authorities may give the Shari'ah a supreme legal authority as the only source of law (hardly any country does that today), while individuals may apply Islamic law in their personal life. In this latter realm the Shari'ah as a moral code for Muslims is fully functional and is better understood as 'the collective ethical subconscious' of the Muslim community.

Medical Ethics as put together by the First International Conference on Islamic Medicine in 1981 addressed the following characteristics:

1. A thorough definition and understanding of the Islamic Medical Profession
2. The acceptable characteristics of the physician
3. The trust between physicians and their patients
4. The Sanctity of Human Life
5. Responsibility and Liability of the physician in providing care
6. Physicians and their roles in the society
7. Physicians approach to modern biomedical advances
8. The effect of innovations in Medical Education

It is optimistic that if these assumptions are followed appropriately, it will lead to heavy dependence upon revelation and no role for human deduction. It is a paramount belief that God's commands are purposeful and human reason in dependence upon revelation can discern rules and apply. The Islamic bioethics or the Islamic medical ethics mentioned that God's command will extend to all areas of life and every field of action since God's will is purposeful. These two tendencies or assumptions gave birth to the richness of Islamic legal thought and ethical reflection in the development of usul-ul-fiqh. Usul-ul-fiqh is the science that identifies the sources of fiqh-law and also lays down rules for weighing these sources against each other in cases of conflict. The sources of Islamic fiqh-law can be divided into material sources and formal sources. According to the Islamic Code of Ethics, the four major schools of law in Sunni Islam agree upon the material sources being the Qur'an and the Sunnah and the other sources being Ijma and Qiyas. The Qur'an is the Muslim holy book and held to be the literal word of God transmitted through the angel Gabriel to the Prophet Muhammad (pbuh) over a period of 23 years. The Sunnah refers to all narrations from the Prophet (pbuh), his acts, sayings, and whatever tactically approved during his time. It is another source of shar'iah along with the holy Qur'an. It aids in understanding the Qur'an. Ijma is to determine and or agreed on something. It is the derivative of Umma – Muslim community. Qiyas literally means 'to compare' or 'to measure', and in practice is juristic reasoning by analogy. In application it stands for applying a certain ruling from an established case if the predisposing conditions, which led to the ruling in the first case, apply to a second case. The four schools of thought Hanafi, Shafi'i, Maliki and Hanbali named after famed jurists developed systems for weighing evidences in the

Shari'ah and promulgated their views in regulating all aspects of a Muslim's public and private lives. They systemized the Islamic law into a comprehensive rational system that covered all possible situations.

The other type of discourse in Islamic Medical Ethics relates to Islamic law, Shari'ah. The Shari'ah is not only a source of law but assigns moral values to actions in Islam. Therefore discussions about medical ethics in Islam are referred to it. This second type of literature aims to define the Islamic stance on biomedical issues ranging from abortion and reproductive technologies to gender relations and end-of-life issues that are further discussed.

Culturally and religious competent healthcare system has so far proven to be the right approach to care considering the diverse nature of current patients. Research shows the healthcare facility that provides culturally and linguistically appropriate healthcare services have the potential to reduce racial and ethnic disparities. A lack of cultural sensitivity by healthcare providers compromises quality of care. Cultural competence training for healthcare providers, the use of linguistically and culturally appropriate health education materials could go along way to improve patient satisfaction and the healthcare industry as a whole. Further, culturally competency in healthcare delivery is expected to fully emerge if the racial and ethnic mix of the workforce is representative of the local population.

Islam Medical Ethics & Quotations from the Holy Qur'an

Below are excerpts derived from the Holy Qur'an that are commonly used in judgments relating to health and healthcare provision and examples of the ways in which such teachings may be applied in healthcare services to the Muslim patients:

Genetic Manipulation, Assisted Conception, & Adoption
"We (God) created Man in the most perfect form." (Qur'an, 95:4)

This indicates that every human life has its own intrinsic values, and that humans however also have the capacity for autonomy and self-determination, and that pursuing a course of action lies in our moral decision therefore, care must be taken to ensure that Islamic principles are adhered to. A complete knowledge of one's pedigree is a fundamental human right, therefore, only somatic cell lines in genetic materials should be transplanted as that will not compromise parental integrity.

"Know your genealogy and respect your blood ties."

Children have the right to be born through a valid union (marriage) and to know their parentage fully. Therefore artificial insemination and in vitro fertilization are allowed or only if the sperm from the woman's spouse is used.

"Call the adoptive children by the name of their father."

Adoption is generally frowned on in Muslim culture since the process involves the transfer of parental rights to the adoptive parents. Fostering is however positively encouraged since no similar transfer of parentage occurs. In either case, the surname of the real father should be retained.

Prenatal Screening and Termination of Pregnancy

"Each of you will have had his created existence brought together in his mother's womb, as a drop (*nutfa*) for forty days, then a leech like clot (*alaqa*) for the same period, then a piece of flesh (*mudga*) for the same period, after which God sends the angel to blow the spirit (*ruh*) into him."

On the basis of this text many Muslims conclude that fetal ensoulment occurs 120 days post-conception and that is an important consideration in discussions regarding termination of pregnancy. First trimester chorionic villous biopsy (performed before ensoulment) and advances in therapeutic fetal medicine may in time lead to a greater willingness to engage in genetic counseling and prenatal screening. An existing life, with its responsibilities and ties, takes preference over a developing one. The Islamic medical ethics agree that when continuation of pregnancy places a mother's life in danger then all Muslim authorities agree that termination of pregnancy is justified. Termination for any other reason is strongly and consistently discouraged, particularly after ensoulment has occurred.

The Qur'an indicates: "Whosoever takes a human life, for other than murder or corruption in the earth, it is as if he has taken the life of all of mankind?" (Qur'an, 5:32)

No one is authorized deliberately to end life, whether one's own or that of another human being. Saving life is encouraged, and reducing suffering with analgesia is however acceptable, even if, in the process, death is hastened. This rule is based on the central teaching that "actions are to be judged by their intentions". Withdrawal of food and drink to hasten death is therefore not allowed.

A US based Muslim philosopher, Rahman, expressed the view that relentless artificial prolongation of life is not in keeping with Islamic ethos unless there is evidence that a reasonable quality of life would result. The majority of Muslim authorities will consider "brain stem" death acceptable grounds to discontinue life support therapy, however, three independent physicians, of whom at least one must be a neurologist, should however make the diagnosis. It is argued that death criteria which remain true to the essence of the Semitic traditions uphold that the point at which the soul departs the body be identified and used for end of life decisions.

Islam and Healthcare Access

Islam regards access to healthcare as one of the fundamental rights of individuals. Therefore, patients have to have the right of care, and the right to make a decision in the course of healthcare utilization. At times, physicians have to decide for patients in light of available knowledge and experience. Furthermore, Muslim physicians derive their conclusion from rules of Islamic laws (Shar'iah) and Islamic medical ethics. The following are some of the guiding principles in Islamic healthcare delivery.

- The first main principle of Islamic Medicine is the emphasis on the sanctity of human life and the safety of mankind derived from the Holy Qur'an.
- The second main principle is the emphasis on seeking a cure and this is derived from the saying of Prophet Muhammad (PBUH) on seeking remedy whenever we fall sick.
- We are encouraged to seek treatment as there is no disease that God has created except that He also has created its treatment.

Thus, when a Muslim physician is making a decision about patient care, that decision should be in the best interest of the patient, whether Muslim or non-Muslim. Further, that decision should not only be based on the physician's own knowledge and experience, but as Muslims, they have to consider the Islamic teaching in regards to the situation and not imposing their religious views on the patient.

All patients irrespective of their faith should be treated with human dignity and respect. Muslim physicians are advised to treat all patients with love and care as if they are members of their own family. Healthcare providers are advised to familiarize themselves with the basic teachings of Islam and Islamic moral values. It is easier for healthcare providers to deal with the patient if they understand their faith, values and culture. These are some of the specific guidelines for healthcare providers of other faith/traditions for caring for their Muslim patients as specified by the Islamic code of medical ethics:

- Muslim patients should be identified, if possible, as Muslim (or with the religion Islam) in the registration information so as to pre-vent any mistakes happening unintentionally in terms of violating dietary rules or privacy.

- Provide Muslim patients islamically slaughtered (Dhabiha) meat. Muslim patients should not be served any pork, pork products or alcohol in their meals. A Muslim patient's family may be allowed to bring food from home, as long as it is meeting the patient's dietary restrictions. If possible, provide accommodation for Muslim patients to perform Islamic prayers.

- Take time to explain test procedures and treatment, and that some of the more recently immigrated Muslims may require the services of translators. Patients consent must always be sorted

- Always examine a female patient in the presence of another female (chaperon) or a female relative (except in medical emergencies). Especially for labor and delivery, if the patient's obstetrician is unavailable and upon her request, provide a female healthcare provider, if feasible. Her husband is encouraged to be present during the delivery.

- After the death of a Muslim patient in a health care facility, allow the family and Imam to arrange for preparing the dead body for burial under Islamic guidelines.

I further presented the Islamic Medical Association of North America's (IMANA) Islamic ethical approach. Their mission is to provide a forum and resources for Muslim physicians and other healthcare professionals, to promote a greater awareness of Islamic medical ethics and values among Muslims and the community at large, to provide humanitarian and medical relief and to be an advocate in health care policy; to promote better understanding and appreciation of the principles of Islamic medicine and to encourage professional interactions amongst Muslim physicians. Their contributions to Islamic medical ethics are as follows:

Responsibility of Muslim Physicians Towards Human Life

Muslims believe that God is the Creator of life and life is a gift from Him. Muslims believe that all life is sacred and must be protected. The respect for life in Islam is common for all humans, irrespective of gender, age, race, color, faith, ethnic origin or financial status. IMANA holds the position that biological life begins at conception while human life begins when ensoulment takes place. A verse from the The Qur'an indicates:

> **"Man, We did create from a quintessence (of clay); Then We placed him as (a drop of) sperm in a place of rest, firmly fixed; Then We made the sperm into a clot of congealed blood; then of that clot We made a (fetus) lump; then we made out of that lump bones and clothed the bones with flesh; then we developed out of it another creature. So blessed be to God, the best to create!" (Qur'an, 23:12)**

Ensoulment (newly created soul) is believed to occur at 40 or 120 days after fertilization, according to different schools of thought. The right of the human fetus in Islam is similar to the rights of a mature human being, including the right to life, the right to inheritance, the right of compensation when injured by willful acts and the right to penalize assailants.

IMANA extends the principles of medical ethics to the patient in a vegetative state. Until the death has been declared, the patient in a vegetative state is considered a living person and has all the rights of a living person.

Assisted Reproductive Technologies and Surrogacy

Imana believes that infertility is a disease and desire for a cure by an infertile couple is natural. However, in Islam, for an action to be permissible all means of achieving that actions are also to be pure. They believe in the sanctity of marriage and the importance of preserving lineage.

The Qur'an indicates: And God has made for you mates (and companions) of your own nature, and made for you, out of them, sons and daughters and grandchildren, and provided for you sustenance of the best: will they then believe in vain things, and be ungrateful for God's favors?

The Qur'an indicates: It is He who has created man from water: then has He established relationships of lineage and marriage: for thy Lord has power (over all things).

Based on these Qur'anic guidelines, IMANA holds the following positions:

- All forms of assisted reproductive technologies (ART) are permissible between husband and wife during the span of their marriage using the husband's sperm and the wife's ovaries and uterus. No third party involvement is allowed.

- They believe in the sanctity of marriage and that the death of the husband terminates the marriage contract on earth, thus frozen sperm from a deceased husband cannot be used to impregnate his widow.

IMANA understands that certain organs may fail in the human body while the rest of the body may still be functional. The current state of medical knowledge holds the view with scientific proof that, if the diseased organs are replaced by healthy organs and if accepted, the body machine can continue to function rather than die because of one diseased organ. Islam instructs all Muslims to save life. Thus, on this basis, transplantation in general, both giving and receiving organs, is allowed for the purpose of saving life, and the following guidelines apply:

- The medical need has to be defined.
- The possible benefit to the patient has to be defined.
- Consent from the donor as well as the recipient must be obtained.
- There should be no sale of organs by any party.
- No financial incentive to the donor or his relatives for giving his organs, but a voluntary gift may be permitted. On the

other hand, there should be no cost to the family of the donor for removing the organ.

- Any permanent harm to the donor must be avoided.
- Transplants of sex organs (testicles or ovaries) that would violate the sanctity of marriage are forbidden.
- Cadaver donation is permitted but only if specifically mentioned in that persons will or in driving license.
- Blood Transfusion is permissible. Giving blood to or receiving blood from people of other faiths is permissible.

Religious Belief and Healthcare Utilization

Presented below are brief points regarding healthcare for patients from various religions that health care providers should keep in mind when caring for them. Healthcare providers are advised to further contact various religious/spiritual leaders to learn how religious/cultural values may be pertinent to a hospital stay regarding personal needs, interaction with staff, and decisions about treatment. They are further advised to be familiar with practical points for the following patients:

1. Buddhist patients highly require peace and quiet atmosphere for meditation.
2. They are culturally based concern about modesty with regards to treatment from opposite sex provider.

3. A majority of them are strictly vegetarian and may reject medications produced from animal products.

4. Some Buddhists may express strong, *culturally* based concerns about modesty, for instance, regarding treatment by someone of the opposite sex.

5. Clinicians should be very specific in the discussion of the use of any drug that may affect awareness, however it should be noted that moderate use of analgesics might actually enable a patient who is struggling with pain to achieve *greater* concentration and "mindfulness" under the circumstances.

6. In end-of-life care, Buddhists may be very concerned about safeguarding their awareness/consciousness. Clarification of the patient's wishes about the use of analgesics in the days and hours before death is strategically important for developing an ethical pain management plan.

7. As patients approach death, medical and nursing staff should refrain from actions that might disturb their concentration or meditation in preparation for dying.

In terms of catholic patients, healthcare providers must be aware that sacraments and blessings by a catholic priest are highly important, especially before surgery or whenever there is a perceived risk of death and If a patient is near death, there may be an urgent request for a catholic priest to offer special prayers "Sacrament of the Sick" and periodic prayers may be demanded. Further, Patients may request Holy Communion (Eucharist) prior to surgery. While a Catholic priest or Eucharistic Minister would typically offer such a patient only a tiny portion of a wafer, patents that are NPO (to have nothing by mouth) should have this request approved by the care team as medically safe. Patients may also have moral questions about treatment decisions, often about the withholding/withdrawing of life-sustaining treatment. Catholic teaching does not generally require any treatment considered "extraordinary means," but a priest may offer authoritative guidance in specific situations.

Healthcare providers should also be aware that Hindu patients may express strong, culturally based concerns about modesty, especially regarding treatment by someone of the opposite sex. Hindus are often strictly vegetarian and may not consume any meat or animal by-products. For such patients, even medications that are produced using animals are likely to be problematic. It should be noted that Hindu family may request that there be constant attendance of the deceased's body, and a family member or representative may wish to accompany the body constantly. A provision should be made to accommodate that.

The Jehovah's Witness patients uphold strict prohibition against receiving blood (that is: red blood cells, white blood cells, platelets, or plasma), be it by transfusion, in medication containing or manufactured using blood products, or in food. Jehovah's Witnesses are noted for cooperating well with health care providers to seek all possible options for treatment that do not conflict with religious concerns. Jehovah's Witnesses are generally well informed of their rights, options for treatment and the consequences of refusal of transfusion. They may wish to discuss aspects of treatment with Elders of the Witness community or consult the Jehovah's Witness Hospital Liaison Committee, who will act in an advisory and intermediary capacity. It is very common for adults to carry at all times a card stating religiously based directives for treatment without blood, and health providers are advised to be aware of that.

The Jewish patients may strictly observe a rule not to "work" on the Sabbath day (from sundown on Friday until sundown on Saturday) or on religious holidays. If so, this religious injunction against "work" which includes prohibitions against using certain tools, and that medical procedures should not be scheduled during the Sabbath or religious holidays (unless they are life-saving), nor should hospital discharges be planned during such times without the consent of the patient. Jewish patients often request a special "Kosher" diet, in accordance with religious laws that govern the methods of preparation of certain foods. Furthermore, some Jewish patients may have culturally-based concerns about modesty with regards to treatment by someone of the opposite sex. Jewish religious laws pose a complex set of restrictions that can affect medical decisions, and patients or family members may request to speak with a rabbi to determine the moral propriety of any particular decision.

Muslim patients may express strong, religiously/culturally-based concerns about modesty, especially regarding treatment by someone of the opposite sex. A Muslim woman may need to cover her body completely and should always be given time and opportunity to do so before anyone enters her room. Muslims may specifically request a diet in accordance with religious laws for "Halal" food, though many Muslims simply opt for a vegetarian diet as a quiet way to avoid religious prohibitions against such things as pork products or gelatin. Forbidden foods are referred to as "Haram." Further, Muslim dietary regulation can affect patients' use of medications; especially drugs that have porcine origins or that contain gelatin or alcohol. The dietary prohibition against alcohol has occasionally raised questions about Muslims' use of alcohol-based hand rubs in the hospital, but such hand rubs should not ultimately prove problematic, because they do not have an intoxicating effect and are used for potentially life-saving hygiene. When in doubt about Muslims' preference in healthcare utilization, the advice of an Imam should be sought.

After Muslim patient has died, the family may request to wash the body and position the bed to face Mecca, and providers must be aware that burial takes place soon as possible. During the thirty-day month of Ramadan, Muslims do refrain from food and drink from dawn until sundown, as such, physicians must determine whether it is medically appropriate to fast while in the hospital, and if so, arrange for appropriate meals for the patient.

Healthcare providers must also be aware of Pentecostal patients' healthcare preferences. They may require a noise-free environment to offer prayers, and may speak in "tongues" (glossolalia). Patients or family members may request a relatively large number of people be allowed in the patient's room for prayers. Healthcare providers should endeavor to accommodate requests of this nature.

Respect for a patient's autonomy requires procurement of informed consent before any medical intervention. This is fundamental to good medical practice. Health professional may not override a valid and applicable advance refusal of treatment. A mentally competent individual has an absolute right to refuse consent for medical treatment, for any reason, even when this may lead to his or her own death. Healthcare providers should familiarize themselves with the concerns of patients from different religious faith. It is always advisable to work with the patient, family members or spiritual leaders to come up with the best solution for them.

Discussions

Ethics is the moral principle act of conducting oneself appropriately (Forester-Miller and Rueinstein, 1992). I have presented ethical approaches by several institutions such as, healthcare, colleges, and international agencies. The codes of ethics guiding institutions are unique but the bottom line of each one of them is to reduce if not eliminate harm to mankind. We have seen how the National Commission for the Protection of Human Subjects of Biomedical and

Behavioral Research grappled with some of the most difficult issues facing researchers and society: When, if ever, is it ethical to do research on children, or on people with mental problems? Furthermore, at international level, I referred to the Belmont Report of 1978, where the commissioners' outline emphasized on the respect for persons, beneficence, and justice as the three items that should govern the conduct of research with human beings. These three principles, they believed, are generally accepted in our cultural tradition and can serve as basic justifications for the many particular ethical prescriptions and evaluations of human action (Belmont Report, 1978).

References to natural law theory are prominent in the works of Catholic theologians and writers as I pointed out earlier. They see natural law as ultimately derived from God but knowable through the efforts of human beings. The influence of natural law theory can be seen in various health issues such as the one on human stem cell research.

The issue of confidentiality is another area that warrants attention in health and other institutions. Research relating to individuals and groups may involve the collection and storage of information that, if disclosed to third parties, could cause harm or distress. Investigators are advised to protect the confidentiality of patients at all cost. During the process of obtaining informed consent, the investigator should inform the prospective subjects in research about the precautions that will be taken to protect confidentiality. Additionally, patients have the right to expect that their physicians and other healthcare professionals will hold all information about them in strict confidence and disclose it only to those who need, and or have a legal right to the information. Treating physicians and their assistance should not disclose any identifying information about patients to an investigator unless each patient has given consent to such disclosure, and unless an ethical review committee has approved such disclosure.

Islamic medical ethics investigators call for a renewed interest in describing the varied ethical constructs of specific populations. The rise of bioethical issues such as abortion, patient-doctor gender relationship, end-of-life care, euthanasia and many more yearn for exploration and enlightenment in Islamic medical ethics. As pointed out earlier, because of innovations in medicine and patient diversity in some advanced countries, the cultural background of both the clinician and the patient influences the care given. As such, a cultural competence increased focus in medical education programs in order to provide effective and efficient clinical care to patients from diverse backgrounds should be encouraged or be made mandatory. Healthcare providers should be mindful of blood transfusion to patients from various religious sects, especially the Jehovah's Witnesses.

Blood transfusion, their religious demand must be respected at all times, and healthcare providers are advised to be aware when caring for patients from various religious traditions. If possible healthcare providers should encourage patients and family members to interpret how religious/cultural values may be pertinent to a hospital stay with regards to personal needs, interaction with staff, and decisions about treatment. They may also contact the chaplains of department of pastoral care for guidance as religious traditions tend to be complex and long.

Healthcare providers are highly advised to adhere to the principles of patient informed consent in providing care. There are negligent-related prices to pay. Healthcare providers should be reminded consistently regarding patient informed consent or be provided with workshops to facilitate that. Study indicated that healthcare entities would see a tremendous reduction in malpractice related costs, and increase their competitive edge if the principles of informed consent are adhered to. Neglecting it, however, poses a serious repercussion, and that healthcare providers are advised to come up with strategies of addressing informed consent issues, as doing so will help avoid several penalties thereby boosting competitive edge.

References

American Medical Association (AMA) (2016). Principles of Medical Ethics. Retrieved from https://www.ama-assn.org/about-us/code-medical-e.thics/principles- medical-ethics.

Aroskar, M., (1998). Administrative Ethics: Perspectives on Patients and Community-Based Care. *Online Journal of Issues in Nursing* Vol. 3, No. 3. Retrieved from: www.nursingworld.org/MainMenuCategories/ANAMarketplace/ANA Periodicals/OJIN/TableofContents/Vol31998/No3Dec1998/AdministrativeEthicsPerspectives.aspx

ABA Commission on law and aging. (2007). Surrogate consent in the absence of an advance directive. Chicago: *American Bar Association*.

Ali YA (1938) *The meaning of the Glorious Quran 95:4*. (Dar al-Kitab, Cairo) (Translation modified).

American College of Rheumatology (ARC, 2015). Education – Treatment – Research. Retrieved from http://www.rheumatology. org/about_us

American counseling association (2005). Code of ethics. *Alexandra*, VA.

Anwar M. (2000). Muslims in Britain: demographic and socio-economic position. In *Caring for Muslim patients*. Sheikh A, Gatorade AR (Radcliffe, Oxford), pp. 3–16

Appelbaum P. (2007). Assessment of patients' competence to consent to treatment. *The New England Journal of Medicine*. Vol. 357:1634-1640.

Australian Medical Association (AMA) (1996). *Code of Ethics*.

Blanco F. and Suresh G. (2005). Ensuring accurate knowledge of prematurity outcomes for prenatal counseling. *Division of Neonatology, Department of Pediatrics*, Vermont Children's Hospital, University of Vermont, Burlington, Vermont, USA.

Bayles M. (1981). Professional ethics. Belmont, CA: *Wada worth Publishing*.

Barry M. (2000). Involving patients in medical decisions: How can physicians do better? *Journal of American Medical Association*.

Belmont Report (1978). *National commission for the protection of human rights*

Berg A. and Braddock C. (2001). Informed consent: *Legal theory and clinical practice.* New York: Oxford University Press.

Brody J. and Scherer D. (2006). Family and physician influence on asthma research participation decisions for adolescents: the effects of adolescent gender and research risk. *Center for Family and Adolescent Research, Oregon Research Institute.*

Browning J. (2008). Pastoral Care and Education. *The University of Pennsylvania Health System.* Penn Medicine.

Bruhn J. & Handerson G. (1991). Values in health care: *Choices and conflicts.* Springfield IL: Charles C. Thomas, Publisher.

Chaplain John Ehman J. (2012). Religious Diversity: Practical Points for Health Care Providers.

Carpenter W. and Lahti A. (2000). Decisional capacity for informed consent in Schizophrenia Research. *General Psychiatry* 57:533-538.

Chima S. (2013). Evaluating the quality of informed consent and contemporary clinical practices by medical doctors in South Africa: *An empirical study. BMC Medical Ethics,* 14(1). Doi: 10.1186/1472-6939-14-S1-S3.

Coulson N. (2016). Shari'ah. *In Encyclopedia Britannica.* Retrieved from http:/www.britannica.com/topic/Shariah.

Darr K. (1997). Ethics in health service management. 3rd Ed. Baltimore, MD: *Health Profession Press.*

Davis T., Berkel H., and Holcombe R. (1998). Informed Consent for Clinical Trials: *a Comparative Study of Standard Versus Simplified Forms.* Oxford Journal of Medicine and Health, National Cancer Institute. Vol. 90, issue 9. 668-674. Doi: 10.1093/jnci/90.9668.

Dixon J. and Smalley M. (1981). *The Surgical/ethical challenge.* 246(27): 2471-2.

Dunn B. and Roberts W. (2005). Emerging findings in ethics of schizophrenia research. *Department of Psychiatry,* University of California at San Diego, USA.

Dunn L., Nowrangi M., and Jeste D. (2006). Assessing decisional capacity for clinical research or treatment: A Review of Instruments. *AM J Psychiatry,* 163: 1323 -1334.

Edge S. & Groves R. (1994). The Ethics of Health Care: A guide to clinical practice. Albany, NY: Delmar Publishers, Inc.

Effa-Heap G. (2004). Blood transfusion: implications of treating a Jehovah's Witness patient.

Etchells E. and Katz M. (1997). Accuracy of clinical impressions and mini-mental state exam scores for assessing capacity to consent to major medical treatment. *Psychosomatics;* 38:239-245.

Farnsworth M. (1990). Competency evaluations in a general hospital. *Psychosomatics* Vlo. 31: 70-76.

Fitten J. and Hamman C. (1990). Assessing treatment decision-making capacity in elderly nursing home residents. *Geriatric Soc:* 38, 1097-1104.

Fisher S. and Alexandra D. (1983). The social organization of doctor-patient communication. Washington DC: *Center for Applied Linguistics.*

Forester H. and Davis T. (1996). A practitioner's guide to ethical decision making. *American Counseling Association.*

Forester-Miller H and Rubenstein L. (1992). Group counseling: *Ethics and professional Issues.* Denver, CO: Love Publishing Co.

Genuis, S. J., & Lipp, C. (2013). Ethical Diversity and the Role of Conscience in Clinical Medicine. *International Journal of Family Medicine, 2013,* 587541. http://doi.org/10.1155/2013/587541

Getrad A. (2000). Medical ethics and Islam: principles and practice Department of Primary Care and General Practice, Imperial College School of Medicine, London, UK

Goodwin P. and Lair T. (1995). Decision-making incapacity among nursing home residents. NMS Survey. *Behavior Science Law,* 13:405-414.

Haas J. and Malouf L. (1989). Keeping up the good work: A practitioner's guide to mental health ethics. Sarasota FL: *Professional Resource Exchange, Inc.*

Hamida F. (1998). Islam and bioethics. *European Network of Scientific Co-operation on Medicine and Human Rights. The human rights, ethical and moral dimensions of health care.* (Council of Europe Publishing, Strasbourg), p 84.

Hammond K. and Lewis R. (2004). Influence of ethical safeguards on research participation: comparison of perspectives of people with schizophrenia and psychiatrists. *Department of Psychiatry and Behavioral Medicine.*

Henley A, Schott J (1999) *Culture, religion and patient care in a multi-ethnic society.* (Age Concern England, London), pp. 1–30.

Hurdle J, et al., (2007). A Code of Professional Ethical Conduct for the American Medical Informatics Association. *J Am Med Inform Assoc.* 14(4): 391–393. Doi: 10.1197/jamia.M2456.

Janvier A. and Barrington K. (2005). The ethics of neonatal resuscitation at the margins of viability: informed consent and outcomes. *Division of Neonatology, Royal Victoria Hospital,* Montreal, Quebec, Canada.

Josen A. (1990). The new medicine and the old ethics. Cambridge, MA: Harvard University Press.

Katz J. (1984). The silent world of doctor and patient. *New York: Free Press.*

Keenan H. and Doron M. (2005). Comparison of mothers' and counselors' perceptions of pre-delivery counseling for extremely premature infants. *Department of Social Medicine,* University of North Carolina.

Kichener K. (1984). Intuition, critical evaluation and ethical principles: *The foundation for ethical decisions in counseling psychology.* Counseling Psychologist, 12(3), 43-55.

Kim H. and Swan J. (2007). Determining when impairment constitutes incapacity for informed consent in schizophrenia research. *Br J Psychiatry;* 191:38-43.

Kim S. and Karlawish J. (2002). Current state of research on decision-making competence of cognitively impaired elderly persons. *Geriatric Psychiatry* 10:151-165.

Kitchener S. (1984). Intuition critical evaluation and ethical principles: *The Foundation for Ethical Decision in Counseling Psychology.* Counseling Psychologist, 12(3), 43-55.

Kogi K, (2002). International Code of Ethics for Occupational Health Professionals (ICOH). Retrieved from http://wwwicohweb.org/core_ethics_eng.pdf.

Levine C. (2008). Taking sides. *Clashing view on bioethicaliIssues.* McGraw-Hill Companies, Inc.

Levey M. (1967). Medical Ethics of Medieval Islam with Special Reference to Al-Ruhawi's 'Practical Ethics of the Physician'. Transactions of the American. Philosophical Society 1967; 57(3): 1–99.

Lillemoen L and Pedersen R (2015). Ethics reflection groups in community health services: an evaluation study. University of Oslo. *BMC Medical Ethics.* Doi:10.1186/s12910-015-0017-9.

Milligan L., Bellamy M., (2004). Continuing education in anesthesia, critical care and pain. Oxford Journals, Medicine and Health. V (4), pp. 35-39. Retrieved from ceaccp.oxfordjournals.org/content/4/2.

McKinnon K. Cournos F. (1989). Rivers in practice: clinicians' assessments of patients' decision-making capacity. *Hosp Community Psychiatry*; 40:1159-1162.

National Institute of Health (NIH, 2001). NIH policy and guidelines on the inclusion of women and minorities as subjects in clinical research. US Department Health and Human Services. Retrieved from grants.nih.gov/grants/fundingwomen_min/ guidelines_amended_10_2001 htm

Nijhawan L et al. (2013). Informed consent: Issues and challenges. Journal of Adv Pharm Technol Res. 4(3): 134–140. Doi: 10.4103/2231-4040.116779

Padela A. (2007). Islamic Medical Ethics: A Primer. Bioethics, vol. 21, 169-178. Journal compilation. Blackwell Publishing Ltd. Oxford, UK.

Parker L. (2007). Informed Consent: *Legal theory and clinical practice*. New York: Oxford University Press.

Rao S. (2008). Informed consent: *An ethical obligation or legal compulsion?* Journal of Cutaneous and Aesthetic Surgery. 1(1): 33-35. Doi: 10.4103/0974-41159.

Rawls J. (1971). A theory of justice. Cambridge MA: *The Belknap Press*.

Rogerson K. (1991). Introduction to ethical theory. Fort Worth, TX: *Holt, Rinehard and Winston, Inc.*

Rosenbaum M. (1982). Ethical problems of group psychotherapy. *Ethics and values Guidebook*: New York Free Press.

Ruth R. and Beauchamp T. (1986). A History and theory of informed consent. New York: *Oxford University Press.*

Saks E., Dunn L., Marshall B. (2002). A new instrument to measure the appreciation component of capacity to consent to research. *Am J Geriatric Psychiatry*, 10:166-174.

Scherer D. and Annett R. (2006)). Family and physician influence on asthma research participation decisions for adolescents: the effects of adolescent gender and research risk. Center for Family and Adolescent Research, *Oregon Research Institute.*

Scott Y. and Kalawish M. (2004). Current State of Research on Decision-Making Competence of Cognitively Impaired Elderly Persons. *American Association for Geriatric Psychiatry.*

Sharman M. and Meert K. (2005). What influences parents' decisions to limit or withdraw life support. Department of Pediatrics, Children's Hospital ofMichigan, *Wayne State University, Detroit.*

Sileo F. & Kopala M. (1993). An A-B-C-D-E Worksheet for promoting beneficence when considering ethical issues. *Counseling and Values*, 37, 89-95.

Singer P. (1994). Rethinking life and death. *The collapse of our traditional Ethics*. New York: St. Martin's Press.

Smith D., and Blumenthal D. (2012). Community Health Workers Support Community-based Participatory Research Ethics. J Health Care Poor and Underserved. file:///Users/amusah/Documents/Community%20Health%20Workers20Support %20Community-based%20Participatory%20Research%20Ethics_.htm

SophiaOmni (2012). Sophia Project. On the supposed right to lie from benevolent motives. *Immanuel Kant*. Retrieved from www.sophiaoomni.org

Stadler H. (1986). Making Hard Choices: Clarifying Controversial ethical issues. *Counseling and Human Development*, 19, 1-10.

The council for international organizations of medical sciences (CIOMS). (2002). *International ethical guidelines for biomedical research involving human subjects.*

Thomas, C., Sage M., Dillenberg J. and Guillory J. (2002). A Code of Ethics for Public Health. *American Public Health Association (APHA)*. Am J Public Health. 92(7): 1057-1059.

Van Hoose W. (1980). Ethics and counseling. *And human development*, 13(1), 1-12.

Vollmann J. and Danker H (2003). Competence of mentally ill patients: *a comparative empirical study*. Psychol Med; 33:1463-1471

Wirshing D., Wirshing W., Marder S., Liberman R., & Mintz J (1998). Informed consent: *assessment of comprehension*. *Am J Psychiatry* 155:1508-1511.